Communications
in Computer and Information Science 2070

Rationale

The CCIS series is devoted to the publication of proceedings of computer science conferences. Its aim is to efficiently disseminate original research results in informatics in printed and electronic form. While the focus is on publication of peer-reviewed full papers presenting mature work, inclusion of reviewed short papers reporting on work in progress is welcome, too. Besides globally relevant meetings with internationally representative program committees guaranteeing a strict peer-reviewing and paper selection process, conferences run by societies or of high regional or national relevance are also considered for publication.

Topics

The topical scope of CCIS spans the entire spectrum of informatics ranging from foundational topics in the theory of computing to information and communications science and technology and a broad variety of interdisciplinary application fields.

Information for Volume Editors and Authors

Publication in CCIS is free of charge. No royalties are paid, however, we offer registered conference participants temporary free access to the online version of the conference proceedings on SpringerLink (http://link.springer.com) by means of an http referrer from the conference website and/or a number of complimentary printed copies, as specified in the official acceptance email of the event.

CCIS proceedings can be published in time for distribution at conferences or as postproceedings, and delivered in the form of printed books and/or electronically as USBs and/or e-content licenses for accessing proceedings at SpringerLink. Furthermore, CCIS proceedings are included in the CCIS electronic book series hosted in the SpringerLink digital library at http://link.springer.com/bookseries/7899. Conferences publishing in CCIS are allowed to use Online Conference Service (OCS) for managing the whole proceedings lifecycle (from submission and reviewing to preparing for publication) free of charge.

Publication process

The language of publication is exclusively English. Authors publishing in CCIS have to sign the Springer CCIS copyright transfer form, however, they are free to use their material published in CCIS for substantially changed, more elaborate subsequent publications elsewhere. For the preparation of the camera-ready papers/files, authors have to strictly adhere to the Springer CCIS Authors' Instructions and are strongly encouraged to use the CCIS LaTeX style files or templates.

Abstracting/Indexing

CCIS is abstracted/indexed in DBLP, Google Scholar, EI-Compendex, Mathematical Reviews, SCImago, Scopus. CCIS volumes are also submitted for the inclusion in ISI Proceedings.

How to start

To start the evaluation of your proposal for inclusion in the CCIS series, please send an e-mail to ccis@springer.com.

Shey-Huei Sheu
Editor

Industrial Engineering and Industrial Management

5th International Conference, IEIM 2024
Nice, France, January 10–12, 2024
Proceedings

 Springer

Editor
Shey-Huei Sheu
Asian University
Taichung, Taiwan

ISSN 1865-0929   ISSN 1865-0937 (electronic)
Communications in Computer and Information Science
ISBN 978-3-031-56372-0   ISBN 978-3-031-56373-7 (eBook)
https://doi.org/10.1007/978-3-031-56373-7

This Springer imprint is published by the registered company Springer Nature Switzerland AG
The registered company address is: Gewerbestrasse 11, 6330 Cham, Switzerland

Paper in this product is recyclable.

Preface

The 5th International Conference on Industrial Engineering and Industrial Management (IEIM 2024) was held successfully in Nice, France from January 10–12, 2024. This conference was sponsored by: Science and Engineering Institute, USA supported by KU Leuven, Belgium; University of Suffolk, UK; Université Le Havre Normandie, France; Royal Holloway, University of London, UK and Universidad de Lima, Perú.

The objective of the conference was to bring together world-class participants and young researchers looking for opportunities for exchanges that cross the traditional discipline boundaries and allow them to resolve challenging multidisciplinary problems, which is only possible at a venue of this nature. Attendees were able to share state-of-the-art developments and cutting-edge technologies in the broad areas of industrial engineering and management.

IEIM 2024 received 71 papers and finally 18 papers were accepted into these proceedings. The type of peer review was double-blind in order to promote transparency and fairness. The conference received great attention from all over the world, with participants from different countries such as Austria, China, Peru, Saudi Arabia, Singapore, etc. According to the accepted papers, the five topics of IEIM were classified as follows: "*Data Analysis and Demand Calculation in Industrial Production*", "*Process Optimization and Intelligence in Green Manufacturing Systems*", "*Lean Manufacturing and Process Optimization*", "*Enterprise Digital Transformation and Business Management*" and "*Modern Logistics Information Systems and Distribution Services*".

The proceedings editors wish to express their deepest gratitude to the Program Committee members, Technical Committee members, reviewers and all contributors for upholding the conference's academic quality.

January 2024

Shey-Huei Sheu

Organization

General Chairs

Stella Sofianopoulou   University of Piraeus, Greece
Roel Leus   KU Leuven, Belgium

Program Chairs

Sheyhuei Sheu   Asia University, Taiwan
Luiz Moutinho   University of Suffolk, UK

Steering Co-chairs

Juan Carlos Quiroz Flores   Universidad de Lima, Peru
Dimitri Lefebvre   Université Le Havre Normandie, France
Adrian E. C. Mondragon   Royal Holloway, University of London, UK

Publicity Chairs

Sadok Turki   University of Lorraine, France
Amr Eltawil   Egypt-Japan University of Science and Technology, Egypt
Hazem W. Marar   Princess Sumaya University for Technology, Jordan

International Technical Committee

Abdulelah Ali   Jazan University, Saudi Arabia
Alberto Flores Pérez   Universidad de Lima, Peru
Charoenchai Khompatraporn   King Mongkut's University of Technology Thonburi, Thailand
Claudia Lizette Garay-Rondero   Tecnológico de Monterrey, Mexico
Danupol Hoonsopon   Chulalongkorn University, Thailand
Deok-Joo Lee   Seoul National University, Republic of Korea

Eugene Khmelnitsky	Tel Aviv University, Israel
Gábor Princz	University of Applied Sciences Wiener Neustadt, Austria
Ghazi Mustafa Magableh	Yarmouk University, Jordan
Hugo d'Albert	Technical University of Munich, Germany
João C. O. Matias	University of Aveiro, Portugal
Jose Antonio Taquia Gutierrez	Universidad de Lima, Peru
José Antonio Velásquez Costa	Peruvian University of Applied Sciences, Peru
Lamia Berrah	Université Savoie Mont Blanc, France
Lina Aboueljinane	Mohammed V University of Rabat, Morocco
Livio Cricelli	Università di Napoli Federico II, Italy
Ludwig Martin	Pforzheim University, Germany
Mahmoud Zeidan Mistarihi	Yarmouk University, Jordan
Malek Almobarek	Alfaisal University, Saudi Arabia
Martin Fidel Collao-Diaz	Universidad de Lima, Peru
Maryam Gallab	ENSMR, Morocco
Matthieu Godichaud	Université de technologie de Troyes, France
Naoufal Sefiani	Abdelmalek Essaadi University, Morocco
Nuno Costa	Instituto Politécnico de Setúbal, Portugal
Nyoman Pujawan	Institut Teknologi Sepuluh Nopember, Indonesia
Omer A. Bafail	King Abdulaziz University, Saudi Arabia
Orhan Korhan	Eastern Mediterranean University, Cyprus
Paolo Trucco	Politecnico di Milano, Italy
Radoslaw Rudek	General Tadeusz Kosciuszko Military University of Land Forces, Poland
Reny Nadlifatin	Institut Teknologi Sepuluh Nopember, Indonesia
Sanjeev Goyal	Thapar University, India
Serena Strazzullo	University of Naples Federico II, Italy
Shiva Abdoli	University of New South Wales, Australia
Sorina Moica	UMFST, Romania
Soroush Avakh Darestani	London Metropolitan University, UK
Supradip Das	Indian Institute of Technology Guwahati, India
Susana Garrido Azevedo	University of Coimbra, Portugal
Tamer Eren	Kirikkale University, Turkey
Xinguo Ming	Shanghai Jiao Tong University, China
Yu-Wang Chen	University of Manchester, UK

Contents

Service Model Based on 5s, Kanban and Standardization of Work to Reduce Rate of Non-service in the Retail Sector

Carlos A. Solano-Ampuero⬩, Alejandro F. Tuanama-Florez⬩,
Martin Fidel Collao-Diaz$^{(\boxtimes)}$ ⬩, and Alberto E. Flores-Perez⬩

Facultad de Ingeniería Industrial, Universidad de Lima, 15023 Lima, Perú
{20181808,20181923}@aloe.ulima.edu.pe, {mcollao,
alflores}@ulima.edu.pe

Abstract. In this study, the relevance of strengthening operations in the Commerce sector is highlighted, since it expands economic opportunities. In addition, within the field of commerce are the retail industries or department stores, which will be the focus of this research. Problems have been identified in the field of trade, such as stock breaks and the lack of efficiency in warehouses and order management. Likewise, it has been observed that the inefficiency of warehouses is related to their size or space, which leads to the accumulation of merchandise and the need to make new investments. Therefore, when analyzing the existence of inefficient stocks and warehouses in retail companies, engineering solutions can be sought to address this problem. This research presents a new solution model that goes beyond common practices observed in previous research, such as the frequent use of 5S. In this case study, the Kanban tool will be additionally implemented and will focus on achieving a correct work standardization. The validation was used the simulation with the Arena tool in the dispatch area, whose results are the undispatched merchandise index was reduced to 3% and the productivity increased to 3,150 units per hour.

Keywords: Retail · 5s · Kanban · Work Standardization · Dispatch · Productivity · Warehouse

1 Introduction

Trade acts as an engine of growth that has led to the creation of better employment opportunities and poverty reduction [1]. In this current study, the significance of enhancing activities within this sector is emphasized, as it increases economic prospects in developing countries, such as Peru [1]. For the mention country, trade contributes 11.55% to the GDP, with a particular focus on the Retail or Department Store industries, which will be the primary focus of this research. It's evident that this sub-activity has generated over 111,462 jobs [2].

Identified issues include stockouts and warehouse efficiency along with order management. Similar problems have been observed in another research; for instance, the

S.-H. Sheu (Ed.): IEIM 2024, CCIS 2070, pp. 1–12, 2024.
https://doi.org/10.1007/978-3-031-56373-7_1

global stockout rate is mentioned at 8%. Another example is warehouse inefficiency, which is proportionate to warehouse size, leading to stock accumulation and new investments. Therefore, by analyzing the presence of inefficient stocks and warehouses in retail companies, new engineering solutions can be developed to address the presented issues. Retail companies in Peru need to enhance efficiency to better satisfy their customers. Thus, a case study employed a proposed implementation using ABC analysis, Holt-Winters, and the Lean Tool: 5S. As a result, the stockout problem significantly reduced additional costs. This research introduces a novel solution model, incorporating the frequent use of 5s from previous studies, with the additional implementation of the Kanban tool and proper work standardization in this case study.

This scientific article is divided into eight sections: introduction, problem analysis, literature review, contribution, results, discussions, conclusions, and references.

2   State of Art

2.1   Productivity Improvement

To determinate the tools to be employed for enhancing productivity in warehouses, information has been gathered from various academic papers. For instance, the implementation of checklists within warehouse areas, the demarcation of pallets parking zones, and the establishment of a sequence for pallet parking led to a 20% increase in storage productivity and a 25% increase a picking productivity [3].

There are also other tools such as Business Process Management (BPM), which involves defining the correct procedures in different areas of the targeted improvement site. Using these tools, the service level was increased from 81.3% to 94.89%, attributed to proper training on the new workflow [4] and ensuring product availability is crucial for retailers, relying on reorder decisions and replenishment efficiency [5]. Finally, the application of SMED, 5S and TPM to increase productivity in food processing in a SME (small and medium-sized company), had a 25% reduction in maintenance times and at the same time a saving in non-productive of USD 137,000 a year [6].

2.2   Using 5S Tool

The 5S methodology enables the optimization of resources by creating a more pleasant working environment through cleanliness and organization. This, in turn, reduces the time spent searching for materials or tools, minimizes the risk of accidents, and enhances productivity [7]. Furthermore, inventory planning is feasible for implementing this methodology, and it can be adapted to various sectors as well [8].

On the other hand, this methodology also enhances worker satisfaction by eliminating wasteful materials enabling them to carry out their tasks more efficiently [9]. It can be concluded that this reduction has led to an 87% decrease in tool search time [7].

2.3   Using Kanban Tool

The Kanban tool has undergone variations in its structure and usage, such as proposed approaches that focus on continuous work in progress, card loops, and card balance control. However, in each reviewed instance, relevant characteristics and common points are present, such as its simplicity and effectiveness. Results demonstrate an optimization of manufacturing time by reducing setup time by 66% and delivery time by 33% [10]. Additionally, Kanban is a tool that enables more active control, as it is present in all activities and can reduce the likelihood of errors [11].

The purpose of the Kanban tool is to maintain a high utilization rate in different workstations while also keeping a low stock of perishable products in the process. In the case of a Peruvian meat company, the Kanban tool was implemented by beginning with a four-week training period for the personnel. This training was divided into several stages, and once properly utilized, it led to a 10% reduction in the time spent on the cutting and packaging processes [12]. For example, a successful Kanban System implementation in an automotive company, along with integrated lean tools, for effective inventory management and problem-solving, applicable to various industries [13].

2.4 Use of Work Standardization

The use of work standardization is a tool that can be applied in any operational area, and it can help reduce time in various processes. For instance, in a Peruvian hardware company, the implementation of work standardization resulted in a 55% reduction in the picking process time, a 57% reduction in the storage process time, and a 67% reduction in the reception process time. All these improvements were confirmed through a simulation using the Arena software [14]. The following Table 1 shows the authors who have contributed the most to identify the tools required for the proposed model.

Table 1. Comparative matrix of the objectives of the problem vs the state of art.

Authors / Causes	Barcode Placement for anchoring dispatch is not situated in the correct area	Operational errors when consolidating goods	Disorder in labeling the destination goods	File overload	Disorder of the materials to be used
[7]			5S		5S
[8]		5S			
[12]	Kanban	Kanban			
[14]		Work Standardization		Work Standardization	
Proposal	Kanban	Work Standardization	5S	Work Standardization	5S

Based on Table 1, the proposed solution has the capacity to address the presented challenges related to the selected items. In this context, the utilization of Kanban, 5S, and Work Standardization will be employed as part of the solution approach.

3   Proposed Model

After presenting these proposals, in a summarized and organized manner, Fig. 1 provides a visualization of the components or phases that this research will include, along with the solutions mentioned in the previous chapter. In other words, the proposed model outlined below is influenced by other research concerning warehouses handling various products, where substantial investment is not necessary to implement improvements.

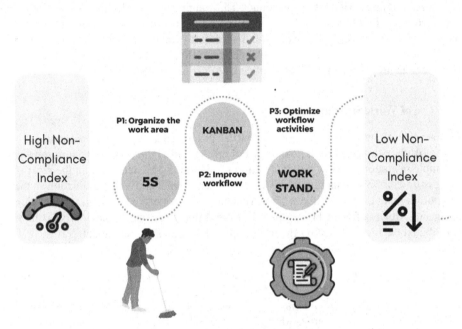

Fig. 1. Proposed improvement model.

3.1   Model Components

Phase 1. Phase 1 includes a series of activities that were previously diagnosed and identified. The primary objective of this phase is to bring organization to the tools utilized within the dispatch area. This is achieved through the systematic implementation of the 5S methodology. The proposed methodology takes into consideration each element of the 5S framework – Sort, Set in order, Shine, Standardize, and Sustain – to effectively categorize, arrange, clean, standardize, and instill a sense of discipline among operators and workers regarding the tools they handle

The overarching aim of Phase 1 is to establish a well-structured environment where tools are systematically arranged, ensuring a clean and clutter-free space. This approach has the potential to greatly reduce process delays that may arise from searching for tools, materials, or products. Moreover, by maintaining this level of organization, the dispatch process becomes more efficient and streamlined, contributing to improved productivity and overall operational efficiency. The ultimate objective is to create an environment where tools are readily accessible, waste is minimized, and workspaces are optimized for maximum effectiveness.

Phase 2. Following the successful organization of Phase 1, the focus shifts to Phase 2, dedicated to enhancing and standardizing the workflow. This critical phase is executed through the strategic implementation of the Kanban tool. The process involves conducting interactive workshops where visual boards with distinct columns are set up, resembling a canvas for collaboration. Within this framework, each operator is equipped with paper or post it notes to identify underlying causes of inefficiencies and propose potential remedies.

The primary goal of Phase 2 is to streamline the workflow, promoting consistency and adherence to standardized processes. By harnessing the power of the Kanban methodology, the workforce is not only encouraged but also empowered to actively engage in the process of problem-solving and process enhancement. This practice fosters a culture of continuous improvement, where iterative adjustments are encouraged, and insights are collectively shared.

The implementation of Kanban not only addresses existing bottlenecks but also aids in the identification of emerging inefficiencies. This iterative approach, coupled with active participation from operators, leads to a refined and optimized workflow that is adaptable to evolving circumstances. Ultimately, the outcome of Phase 2 contributes to enhanced operational efficiency, reduced cycle times, minimized waste, and an overall improvement in the quality of outputs.

Phase 3. The objective is to achieve further workflow optimization by introducing standardized processes. This phase focuses specifically on enhancing the utilization of computers. The central aim is to educate operators about the correct methods of organizing, classifying files, and addressing operational errors. By implementing uniform procedures for computer usage, the intention is to establish a consistent approach that not only increases efficiency but also minimizes errors and discrepancies arising from improper file management. The educational aspect of Phase 3 plays a crucial role in empowering operators to navigate computer systems effectively. This, in turn, contributes to smoother workflows, reduced error rates, and enhanced overall productivity. By adhering to standardized protocols, the risk of operational errors diminishes, resulting in a more streamlined and efficient process. Through these standardized practices, Phase 3 endeavors to create an environment where the entire workflow benefits from improved organization, reduced errors, and optimized computer usage, thereby boosting overall operational effectiveness.

3.2 Model Indicators

In order to measure the progress achieved by the model under consideration, the utilization of the following indicators is put forth in Table 2.

Table 2. Model indicators.

Indicator	Formula	Expectation	Use
Percentage of Compliance for Operations	$\sum$ (Completed Operations/Total Operations)	95%	Verify the fulfillment of each operation
Time taken to search for each tool	Total search time	5 min	Measurement of search time for each tool by each operator
5s Audit	$\sum$ (5S audit score)	85%	Assists in confirming outcomes through the implementation of the 5S methodology
Loading time	Total loading time	0.67 h	Time required for merchandise loading
Goods dispatch	Dispatched Goods/hours	3,300 unit/hour	Merchandise dispatched productivity per hour
Non-Service index	Unattended Goods/ Total Goods	5%	Check the index of unattended cases concerning the undelivered merchandise

4 Validation

4.1 Initial Diagnosis

The company being studied is involved in marketing products across a range of business sectors, encompassing department stores, home improvement and construction, commercial financing, banking, travel, and insurance [15]. The specific focus of this investigation centers on the dispatch area, within which a notable issue has been identified concerning the merchandise dispatch process within the company's warehouse operations. A key challenge revealed is the significant delay experienced by operators in locating their necessary work materials. Additionally, there are concerns related to the proper stacking and arrangement of goods on pallets during the dispatch process. These operational challenges highlight the critical need to rectify inefficiencies within this aspect of the company's overall operations to ensure smoother and more effective processes.

In a detailed time, usage analysis spanning 162 min, key insights have emerged. A significant portion of this timeframe, around 40%, is spent on meticulously reviewing checklists before initiating loading tasks. This extended checklist review is primarily due to recurring issues with pallet handlers neglecting to secure merchandise on the dispatch dock, leading to subsequent waiting times for validation procedures. Another substantial delay, accounting for 34%, is attributed to the improper consolidation of goods, stemming from inadequate checklist oversight, resulting in the accumulation of unused items. Furthermore, an additional delay arises during system closure, as previous shift workers frequently leave files on their computers due to a lack of standardized shutdown protocols. This oversight contributes to avoidable bottlenecks. The analysis has culminated in the construction of a problem tree (Fig. 2), systematically pinpointing underlying causes and pivotal root causes, guiding the implementation of strategies and tools to effectively address these complex challenges.

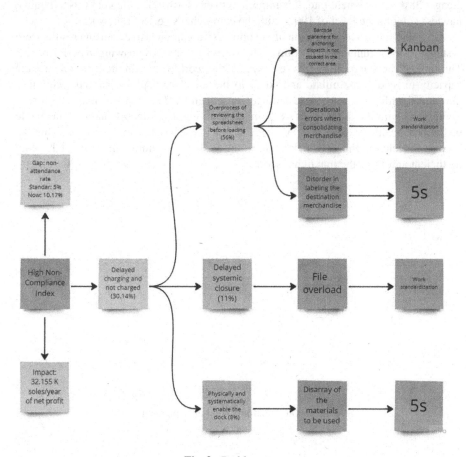

Fig. 2. Problem tree

4.2 Desing of the Validation and Comparison with the Initial Diagnosis

The case study was scheduled for June and confirmed in the beginning of July. An initial trial of the 5S methodology was performed within the warehouse to authenticate the application of enhancements. The Arena software was used to simulate the Standardized Work and Kanban, enabling the assessment of the resulting metrics.

4.3 Improvement

Pilot Test. In relation to the application of the 5S methodology, an initial audit was undertaken to assess the existing status and pinpoint any deficiencies present within each of the five distinct phases of the 5S approach. As a direct outcome of this audit, an initial score of 19 points out of a possible 50 was attained, consequently reflecting a meager 38% achievement rate. It is important to emphasize that the goal was to attain a flawless compliance level of 100% with the comprehensive 5S framework.

Upon the successful execution of the pilot test, comprehensive data and results were systematically accumulated, unequivocally verifying a marked upswing in performance. Throughout the duration of this test period, the workforce exhibited the remarkable capacity to readily assimilate and adapt to the novel methodologies and techniques introduced within the warehouse setting. Consequently, the result was nothing short of impressive, with a total score of 48 points, signifying a substantial and commendable 96.7% enhancement when juxtaposed against the initial baseline measurement. This can be seen reflected in Fig. 3, which explains the 5s, having a minimum score of 1, being inefficient and 10 as the maximum score.

Fig. 3. 5S audition before and after implementation.

Improvement. The simulation encompassed the complete process of dispatching merchandise, encompassing a comprehensive assessment of its various facets. Notably, one of the prime time-consuming activities was identified as the retrieval of PDTs or portable data terminals, pivotal for the collection of products. Given the inherent fragility of these products, ensuring careful handling to prevent any damage to the shipped orders emerged as a crucial concern. Additionally, it became evident that certain operators necessitated precise guidelines pertaining to the proper handling of specific product types, the appropriate stacking procedures, and the requisite load-bearing capacity of individual pallets. To effectively address these challenges, the strategic decision was made to integrate the concept of Standard Work into the operational framework. This strategic move bore the potential to curtail the time invested in an activity that yielded minimal added value, whilst concurrently facilitating an accurate tally of package quantities. It is of noteworthy significance that the simulation process was executed utilizing the Arena software, serving as the quintessential tool for conducting these intricate simulations. A visual representation of this simulation is provided in Figure 4, which offers a tangible depiction of the process dynamics and variables at play.

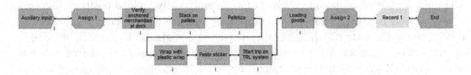

Fig. 4. Simulation of procedure.

To ascertain the appropriate number of replications for the research effort, an initial allocation of 30 replications was deemed fitting. A desired half-width of 1 was set as a benchmark, and subsequently, a total of 5 trial runs were conducted. Concerning the fifth iteration of the simulation, a half-width of 1.0 was yielded, accompanied by a confidence interval of [38.53–40.55].

The simulation optimized activity timings in the dispatch process to reduce warehouse goods dispatch times. Firstly, operation compliance improved from 31% to 96.7%, ensuring almost all activities were fulfilled. Operators' tool search time was halved to 4 min, enhancing the dispatch time. Secondly, loading time dropped from 1.6 to 0.65 h (or 96 to 40 min), boosting dispatched goods productivity from 1,000 to 3,150 units per hour. Dispatch time improved to 0.65 h, indicating positive outcomes. Lastly, the

Table 3. Indicator comparison.

Indicator	Unit	As Is	Target	To be
Percentage of Compliance for Operations	%	31	95	Verify the fulfillment of each operation
Time taken to search for each tool	Minutes	8	5	Measurement of search time for each tool by each operator
5s Audit	%	38	85	Assists in confirming outcomes through the implementation of the 5S methodology
Loading Time	Hours	1.6	0.67	Time required for merchandise loading
Goods Dispatch	Unit/Hours	1,000	3,300	Merchandise dispatched productivity per hour
Non-Service Index	%	10	5	Check the index of unattended cases concerning the undelivered merchandise

simulation highlighted 3% undispached merchandise, with 3,150 items dispatched out of 3,250 scheduled. This summary is detailed in Table 3.

5 Discussion

The implementation of the improvement has unequivocally yielded positive outcomes across various metrics. Notably, the compliance rate for each operation consistently resides within the 66 to 67 percent range, indicating a marked enhancement in meeting operational benchmarks. Additionally, the search time for essential tools during tasks has been significantly curtailed, now spanning between 4.50 to 5 min, signifying a substantial stride in process efficiency. Moreover, the 5S audit has showcased an augmented percentage score, coupled with reduced hours in merchandise dispatch, leading to favorable operational impacts. The productivity surge, elevating merchandise dispatch quantities from 1,000 to 3,150 units per hour, translates into a substantial advantage in managing heightened demand scenarios. Finally, a notable elevation in the non-compliance index underscores the team's improved adaptability and response to unforeseen challenges, fostering a more streamlined and proficient dispatch process. In summation, these improvements have fostered a clear and beneficial impact on pivotal dispatch indicators. From heightened operational compliance and expedited tool access to increased productivity and a sharper response to unexpected scenarios, these enhancements collectively bolster operational efficiency, ultimately amplifying the overarching performance of the company.

6 Conclusions

The synthesis of extensive research underscores that combining 5S, Kanban, and Standardized Work methodologies strategically can augment productivity and mitigate undispached merchandise concerns in retail warehousing. Notably, the study identifies two pivotal subprocesses—pre-loading sheet reviews and post loading inspections. Detailed time analysis reveals a substantial allocation of approximately 162 min for sheet reviews, stemming from haulers' inadequate merchandise anchoring at dispatch bays.

Key issues include over-processing, constituting 56% of inefficiencies, traced to barcode misplacement during pre-loading sheet review, extending loading time to 1.60 h. Kanban rectifies this, slashing loading time to 0.65 h. Integration of 5S and standardized work elevates compliance rates to 96.70% from 31%, reduces tool retrieval time to 4 min, bolsters the 5S audit to 86% from 38%, and boosts productivity from 1,000 to 3,150 units per hour, curbing undispached goods from 10% to 3%. Simulations validate findings, confirming the project's transformative potential.

References

1. World Bank Group.: Commerce. World Bank (2021). https://www.bancomundial.org/es/topic/trade/overview
2. Ministry of Production.: Conjunctural Report - Manufacturing Industry. Results of the Industrial Production Indicator. Lima, Lima, Peru (2022). https://ogeiee.produce.gob.pe/index.php/en/shortcode/estadistica-oee/estadisticas-manufactura
3. Figueroa, R.E., Bautista, G.A., Quiroz, F.J.: Increased productivity of storage and picking processes in a mass consumption warehouse applying Lean Warehousing tools: A Research in Peru. Latin American and Caribbean Consortium of Engineering Institutions (2022). https://doi.org/10.18687/LACCEI2022.1.1.120
4. Palomino, C.J., Camacho, O.R., Macassi, J.I.: Increasing the service level in an industrial supplier company using the Winters Forecasting Method, Lean Warehouse and BPM. Latin American and Caribbean Consortium of Engineering Institutions (2022). https://www-scopus-com.ezproxy.ulima.edu.pe/record/display.uri?eid=2-s2.0-85139984226&origin=resultslist&sort=plff&src=s&sid=d43f97fdc40719e7c7f729a454a45b75&sot=b&sdt=b&s=AUTHORNAME%28Macassi%29&sl=20&sessionSearchId=d43f97fdc40719e7c7f729a454a45b75
5. Marquez, P., Jorge, D., Reis, J.: Using Lean to Improve Operational Performance in a Retail Store and E-Commerce Service: A Portuguese Case Study. MDPI (2022). https://www-scopus-com.ezproxy.ulima.edu.pe/record/display.uri?eid=2-s2.0-85130530450&origin=resultslist&sort=plff&src=s&st1=warehouse+kanban&sid=3b6abad56b8d823eac99cdcccc5c070b&sot=b&sdt=b&sl=31&s=TITLE-ABSKEY%28warehouse+kanban%29&relpos=2&citeCnt=1&searchTerm=
6. Valverde-Curi, H., De-La-Cruz-Angles, A., Cano-Lazarte, M., Alvarez, J. M., Raymundo-Ibañez, C.: Lean management model for waste reduction in the production area of a food processing and preservation SME. In: Paper presented at the ACM International Conference Proceeding Series, 256–260 (2019). https://www-scopus-com.ezproxy.ulima.edu.pe/record/display.uri?eid=2-s2.0-85076710697&origin=resultslist&sort=plf-f&src=s&sid=2f9939a87716ef990e5aae2923b4c904&sot=b&sdt=b&s=TITLE-ABSKEY%28Lean+management+model+for+waste+reduction+in+the+production+area+of+a+food+processing+and+preservation+SME%2109&sl=121&sessionSearchId=2f9939a87716ef990e5aae2923b4c904
7. Velásquez-Costa, J. Impact of the 5S methodology in the optimization of resources in metal mechanical companies. Latin American and Caribbean Consortium of Engineering Institutions (2022). https://www-scopuscom.ezproxy.ulima.edu.pe/record/display.uri?eid=2-s2.0-85139990801&origin=resultslist&sort=plff&src=s&st1=warehouse+5S&sid=2539449f2e08ff67f499aa83489defef&sot=b&sdt=b&sl=27&s=TITLE-ABSKEY%28warehouse+5S%29&relpos=6&citeCnt=0&searchTerm=
8. Mantilla, R.B., Arivilca, L.P., Aparicio, V., Nunura, C.: Inventory management optimization model based on 5S and DDMRP methodologies in commercial SMEs. Latin American and Caribbean Consortium of Engineering Institutions from (2021).https://www-scopuscom.ezproxy.ulima.edu.pe/record/display.uri?eid=2-s2.0-85122041751&origin=resultslist&sort=plff&src=s&st1=warehouse+5S&sid=2539449f2e08ff67f499aa83489defef&sot=b&sdt=b&sl=27&s=TITLE-ABSKEY%28warehouse+5S%29&relpos=12&citeCnt=0&searchTerm=
9. Leon-Enrique, E., Torres-Calvo, V., Collao-Diaz, M., Flores-Perez, A.: Improvement model applying SLP and 5S to increase productivity of storaging process in a SME automotive sector in Peru. Association for Computing Machinery (2022). https://www-scopuscom.ezproxy.ulima.edu.pe/record/display.uri?eid=2-s2.0-85132035327&origin=resultslist&sort=plff&src=s&st1=warehouse+5S&sid=2539449f2e08ff67f499aa83489defef&sot=b&sdt=b&sl=27&s=TITLE-ABSKEY%28warehouse+5S%29&relpos=1&citeCnt=0&searchTerm=

10. Rojas-Benites, S., Castro-Arroyo, A., Viacava, G., Aparicio, V., Del Carpio, C.: Reduction of waste in an SME in the meat sector in peru through a lean manufacturing approach using a model based on 5S, standardization, demand forecasting and kanban. Association for Computing Machinery (2021). https://www-scopus-com.ezproxy.ulima.edu.pe/record/dis play.uri?eid=2-s2.0-85123057474&origin=resultslist&sort=plff&src=s&st1=warehouse+ kanban&sid=f0cc2d8cd70064b043e1dba1f4a4d9dc&sot=b&sdt=b&sl=31&s=TITLE-ABS KEY%28warehouse+kanban%29&relpos=6&citeCnt=1&searchTerm=

11. Canales J.L., Rondinel O.V., Flores P.A., Collao D. M.: Lean model applying JIT, Kanban, and Standardized work to increase the productivity and management in a textile SME. Association for Computing Machinery (2022). https://www-scopuscom.ezproxy.ulima. edu.pe/record/display.uri?eid=2-s2.0-85132043167&origin=resultslist&sort=plff&src=s& sid=537b3a45c9735d25a6bb9d73aa1929b1&sot=b&sdt=b&s=AUTHORNAME%28Rond inel%29&sl=21&sessionSearchId=537b3a45c9735d25a6bb9d73aa1929b1

12. Namay-Zevallos, W., Martinez, M., Soto, C., Salas, R.: A combined demand management and lean tools model at a peruvian meat products company. Association for Computing Machinery (2021). https://www-scopuscom.ezproxy.ulima.edu.pe/record/display.uri?eid=2-s2.0-85123053251&origin=resultslist&sort=plff&src=s&st1=warehouse+kanban&sid=f0c c2d8cd70064b043e1dba1f4a4d9dc&sot=b&sdt=b&sl=31&s=TITLE-ABSKEY%28ware house+kanban%29&relpos=7&citeCnt=0&searchTerm=

13. Ilyiana, T., Ilyin, A.: kanban system as an effective inventory management method for auto-motive component companies. Nova Science Publishers, Inc. (2022). https://www-scopus-com.ezproxy.ulima.edu.pe/record/display.uri?eid=2-s2.0-85137663972&origin=resultslist

14. Campos-Sonco, J., Saavedra-Velasco, V., Quiroz-Flores, J.: Warehouse management model to increase the level of service in peruvian hardware SMEs. Latin American and Caribbean Consortium of Engineering Institutions (2022). https://www-scopuscom.ezproxy.ulima. edu.pe/record/display.uri?eid=2-s2.0-85140006971&origin=resultslist&sort=plff&src=s& st1=warehouse+5S&sid=2539449f2e08ff67f499aa83489defef&sot=b&sdt=b&sl=27&s= TITLE-ABSKEY%28warehouse+5S%29&relpos=4&citeCnt=0&searchTerm=

15. Falabella Group. CIA TALENTS GROUP (s.f.). https://grupociadetalentos.com/falabella/

Demand Forecasting in the Food Equipment Industry Using Predictive Data Analytics Techniques

Ammar Y. Alqahtani[1]([⊠]) [iD] and Anas A. Makki[2] [iD]

[1] Department of Industrial Engineering, Faculty of Engineering, King Abdulaziz University,
Jeddah, Saudi Arabia
aaylqahtan@kau.edu.sa
[2] Department of Industrial Engineering, Faculty of Engineering–Rabigh, King Abdulaziz
University, Jeddah, Saudi Arabia
nhmakki@kau.edu.sa

Abstract. Although food and restaurants are two of the most trending entertainment sectors in Saudi Arabia, where the equipment industry is a massive contributor to the enrichment of this trending environment, a major appliance manufacturer is facing difficulties forecasting the demand in the market. This study aims to study the sales department of one of the biggest appliance traders in the MENA region and forecast their sales for 2021. This study predicted sales using time series methods such as linear progression, exponential smoothing, Seasonal, and Trend decomposition using Loess (STL) and Auto-Regressive Integrated Moving Average (ARIMA). The program used in forecasting and using all these methods was Power BI, including R script commands. The results of the ARIMA method showed higher accuracy and less error than the other methods. The ARIMA method demonstrated an accuracy of 92.16% and a 9.24% error. For more accurate demand forecasting, businesses should consider the market changes for at least two years to understand consumer behavior that might change faster than expected, directly affecting the restaurant industry and the demand for equipment.

Keywords: Predictive data analytics · Demand forecasting · Equipment industry · Restaurants industry · Food industry · Supply chain · Logistics · Appliances manufacturers

1 First Section

Restaurants, cafés, food services, and bars/taverns all fall under the umbrella term "food and beverage industry," which describes establishments that provide raw materials and finished products for the food and drink production process. Manufacturers must customize their goods and services to appeal to various types of customers. The nature of the food goods being easily spoiled, cumbersome to transport, and subject to seasonal fluctuations gives rise to the industry's signature traits. As a result, these traits need careful management [1].

© The Author(s), under exclusive license to Springer Nature Switzerland AG 2024
S.-H. Sheu (Ed.): IEIM 2024, CCIS 2070, pp. 13–29, 2024.
https://doi.org/10.1007/978-3-031-56373-7_2

With a projected CAGR (Compound Annual Growth Rate) of 3.8% between 2017 and 2023, the $9,105 million in 2016 revenue generated by the global commercial cooking equipment market is likely to increase to $11,740 million by 2023. The food service and processing industries could not function without commercial cooking equipment. Various uncooked and cooked foods are prepared using cooking appliances, including ovens and fryers. Equipment of this kind is widely used in commercial kitchens because it facilitates food preparation, improves food safety, and decreases operating and labor costs [2].

Forecasting is a human way of thinking. The basis is that what happened in the past will happen in the future. From this idea, manufacturers and businesspeople adapted this method to determine how much they should produce their products. Demand forecasting refers to predicting the consumer demand that would occur at a specific time in the future [3]. Regardless of the nature of the business, the goal is meeting customer demand, which reflects on the business's profitability. Otherwise, the business might mismanage its recourses. This matches the objective of forecasting to make a rational decision in businesses based on an accurate analysis of future events. Moreover, managers' decisions should be based on demand forecasting to understand the market changes [4].

A well-known appliance manufacturer must adapt to market demand and changes through good planning. This is to increase profitability and reduce cost as the business follows a primitive ordering system that only orders when it has a stock shortage, leading to opportunity cost and some difficulties of sales and procurement that lead to either excessive orders or shortage. The business just started using the moving average method for better coordination. In this paper, different forecasting practices will be performed to determine the most appropriate and accurate forecasting method that would improve the sales planning of the business and minimize inventory costs.

The remainder of this paper is structured as follows. Following this introduction, a literature review is presented covering relevant previous research related to demand forecasting techniques, inventory management approaches, and their applications within the food and restaurant industries. Next, the methodology section details the forecasting methods employed in this study, including quantitative time series techniques and qualitative approaches. It also describes the data and steps taken in the analysis. Subsequently, the results of implementing the various forecasting models on the company's historical sales data are provided. Then, a discussion of the findings and an assessment of each model's accuracy. Finally, the paper concludes with a summary of the key conclusions that can be drawn from the study and opportunities for future work.

2   Literature Review

Demand prediction is crucial for making judgments, considering several elements, and justifying choices. The quantity of stock on hand to absorb demand swings is specified in a demand prediction. Knowing the company's internal workings and market state is crucial. Companies may better satisfy future consumer expectations for goods, stock-keeping units (SKUs), quantities, and facilities with the help of demand forecasting. Eventually, the company can save costs and enhance profits via accurate forecasting, which will build a database to aid decision-makers in achieving goals, formulating strategies, and documenting environmental changes. In addition, it directs the company toward the most effective means of enhancing productivity without compromising the quality of its service to its consumers [5].

Predictions based on any of the quantitative approaches may be improved by including the qualitative (or judging) approach. Executive views, the Delphi approach, sales-force polls, and consumer surveys are four of the most well-known qualitative forecasting methodologies. Sales projections based on the aggregated opinions of executives and industry specialists in marketing, operations, finance, supply chain management, and administration are known as "Executive Opinions." This technique is often used in tandem with a quantitative approach, such as trend extrapolation. In light of their own projections, the management team adjusts the outcome. This method allows for rapid and straightforward forecasting without the need for complex statistical analysis. However, one of the risks of this strategy is groupthink, which may develop when there is excellent cohesion, strong leadership, and isolation inside the group. For instance, when a group of specialists is polled about their predictions for the future using the Delphi Method. The experts avoid holding group meetings to lower the likelihood that agreement is established due to strong personality variables. An independent party summarizes the projections and their supporting justifications before sending them back to the experts with follow-up questions. This process is repeated until everyone agrees on something. If the objective is to make a prediction far into the future, this method will serve well. The method is conducted in a questionnaire style, removing groupthink's drawbacks. In this case, neither a committee nor a vote is necessary. Since the goal is neither agreement nor unanimity, the experts are not swayed in any direction by social pressure to make a particular prediction. However, consensus issues and limited dependability still exist as potential drawbacks of the Delphi technique [6].

For the time series forecasting model, [7] accessed the websites of 56 different businesses trading on India's National Stock Exchange (NSE) across seven different industry sectors (an average of eight companies per sector). They used twenty-three months' worth of training data, from April 2012 to February 2014, to make predictions about the upcoming months' worth of data, and they split their dataset into three different time intervals: six months, from September 2013 to February 2014; twelve months, from March 2013 to February 2014; and eighteen months, from September 2012 to February 2014. The authors created A time series, which is a set of discrete data points gathered at regular intervals. Time series analysis is a crucial subfield of statistics that examines data sets to learn about their properties and aids in forecasting future values of the series by using those features. Methods for analyzing time series include the ARMA model and the ARIMA model, which is derived from the ARMA model.

Several studies in recent years have continued applying and advancing demand forecasting methods across various industries. Dargar et al. [8] employed an ARIMA model optimized through a genetic algorithm for auto parts demand prediction. Their ARIMA-GA approach outperformed ARIMA and other techniques. Tseng et al. [9] combined ARIMA and LSTM networks for stock price forecasting, finding their hybrid model outperformed individual methods. Çelik et al. [10] evaluated STL, ARIMA, and SVM for Turkish steel sector demand, with STL-SVM achieving the best performance. Zhai et al. [11] integrated ARIMA and LSTM models for electric vehicle charging demand forecasts, demonstrating ensemble forecasting benefits. Livieris et al. [12] compared linear regression, SVR, and RNN for olive oil production forecasts in Greece, exhibiting RNN's superior performance. These more contemporary studies continue applying established approaches like ARIMA while exploring techniques such as deep learning and ensemble predictive modeling, reinforcing demand forecasting's ongoing research relevance to supporting strategic business decisions.

Others, like Taylor et al. [13], prefer to represent complicated seasonality using exponential smoothing approaches, which established a model that can handle non-integer periods, high-frequency multiple seasonal patterns, and dual calendar impacts. The data set used in that paper does not include any days that fall on holidays for simplification.

The authors Indrasen et al. [14] have conducted a study to illustrate the benefits of using the ABC analysis in proposing effective criteria for inventory control. They concluded that responding quickly to environmental changes may lead to more promising outcomes. Either a person or, as is recommended here, a decision support system may achieve this by comparing the predicted and actual demand and adjusting the orders accordingly.

In addition, Qin et al. [15] propose two novel hybrid methods for accurately predicting monthly passenger flow in China, one based on seasonal trend decomposition procedures based on loess (STL) and the other on echo state networks (ESN) enhanced by grasshopper optimization algorithm (GOA) and adaptive boosting (Adaboost) framework. According to the proposed methods (STL-GESN, STLAESN), the original passenger flow data are firstly decomposed into seasonal, trend, and remainder components via STL. Then, the improved ESN is adopted to forecast the trend and the remaining components, and the seasonal-naive method is utilized to forecast the seasonal component. Finally, the forecasting results of the three components are summed to obtain the final forecasting of monthly passenger flow. Two passenger flow forecasting applications are carried out to test the viability and scalability of the suggested methods, one using air data and the other using railway data. The experimental results show that STL-GESN and STL-AESN obtain higher prediction accuracy than different forecasting approaches. Application studies also demonstrate that the proposed approaches are the practical choice for passenger flow forecasting.

In addition, one of the most useful applications of statistics is regression analysis, which involves formulating a mathematical model to link dependent and independent variables. Generalized least squares regression is used in three distinct regression models: the Variable-based Degree-Day Model (VBDD), the Linear Regression Model, and the Change-Point model. Likewise, the Artificial Neural Networks (ANN) technique is used in different fields of forecasting building energy use for short- and long-term periods. They provide an attractive way of determining the dependence of energy consumption on occupancy-dependent factors and weather variables [16].

3   Methodology

3.1   Forecasting Methods

The following are some of the forecasting methods that will be used in this study.

Qualitative Forecasting. The executive opinion is a trend forecasting method involving polling a group of well-respected authorities on a specific subject, such as the state of the market for particular securities. Individually, the jury members provide their first evaluations; after reviewing the work of their peers, they collectively make any necessary adjustments to their estimations. Executive opinion juries are helpful because they function as a peer review, highlighting flaws constructively without resorting to litigation [17].

One of the most popular ways to predict future sales is via a market survey, which is conducted to acquire data about the market that cannot be obtained through internal corporate records or publicly available sources. When primary data, such as information gathered from a survey of potential customers, is needed to predict demand, businesses often turn to market research surveys. When this happens, the corporation turns to primary data to determine the best way to bring a new product or product variation to the market. Also, a corporation venturing into uncharted territory will do a market assessment to anticipate future sales. The company must get information from the market or clients directly to make a sales prediction, as the company does not have any historical data. Companies often undertake consumer surveys to learn more about their target audience's spending patterns, opinions, and attitudes. The channel partners' views, intentions, and general industry trends are often polled. The term "market survey" is sometimes misused as a synonym for "market research" or "market analysis." A market survey is a commonly used tool for market researchers. The key benefit of using a market survey approach is that it facilitates the collection of primary data or data that has not previously been collected for any other purpose. Yet, gathering original data may be time-consuming and costly [18].

The sales estimate for a composite is calculated by aggregating the sales projections from sales representatives in the company for their separate territories. Business organizations employ this bottom-up method to improve the accuracy of their forecasts. Salespeople have the most direct contact with consumers, so they are in a prime position to provide insightful feedback that may help businesses increase their revenue. The sales force composite forecast allows the organization to predict the whole market and specific regions. However, if salespeople are the only ones contributing to the company's estimate, they may be too optimistic or gloomy based on their most recent data. As a result, the corporation risks making predictions based only on short-term considerations and ignoring the longer-term macroeconomic situation. As a result, businesses often combine the top-down projection with the sales force composite forecast before settling on the actual estimates. Another possible pitfall of this method is that some agents may under-predict revenues to more easily meet their quota and cash in on the company's incentive for exceeding sales projections. Using historical data and prior predictions, many businesses now widely use scripting computer software. Such software considers the agents' responses and provides a cumulative forecast [19].

The Delphi method is a framework for making predictions based on the responses of a group of experts to a series of questions. After each round of questionnaire distribution, the aggregated, anonymous replies are compiled and presented to the group. In later rounds, the specialists will have the opportunity to modify their responses. The Delphi technique uses repeated questioning and feedback from the group to reach a consensus on the best course of action. The Delphi technique is used to obtain consensus amongst a large group of experts without having to convene everyone in one place. The panel members may speak freely without fear of retaliation since their comments will be kept confidential. Over time, when people's perspectives shift, consensus might be formed. While the Delphi technique does allow for input from a wide range of people, it does not provide the same level of engagement as a face-to-face meeting. The pace of conversation may be slowed by the length of time required for replies. The returned information from the experts could not be helpful in any way [20].

Quantitative Forecasting. Associative and Time Series models are used for making predictions. Forecasters may extrapolate a future value, range of values, or scenario by analyzing historical information. It's crucial for both immediate and far-off endeavors. Quantitative forecasting methods like Time Series and Associative Models are more objective than qualitative methods like the Delphi Technique and market research [21].

The cost of a fixed asset is usually written off using the straight-line depreciation technique throughout its useful life. It has been used when there is no way to predict how an asset will be used in the future. While the straight-line technique is the simplest to use, it does introduce some room for mistakes in the depreciation calculations. Therefore, it is not ideal. The first phase of the straight-line calculation is to ascertain the original cost of the asset that is now a fixed asset. Take the asset's book value and deduct the amount you expect to get for its salvage. Find out how long you can expect an asset to last. It is most convenient to simultaneously apply a universal useful life to all asset types. To calculate the straight-line depreciation rate, divide the expected useful life by one. Multiply the asset's cost by the annual depreciation rate (less salvage value).

Once depreciation expenditure has been computed, a debit entry will be made in the depreciation expense account, and a credit entry will be made in the cumulative depreciation account in the books of account. In this case, the fixed asset account is reduced by the accumulated depreciation account, which is a counter-asset account [22].

3.2   ABC Analysis

The principle upon which the ABC method of inventory control rests is that a relatively small number of items may constitute a relatively small portion of the monetary value of stores. In contrast, a rather large number of items may represent the bulk of total material usage of the entire inventory during the construction process. By adding up how much material goes into making everything, multiply the unit price by the number of items made from that material. Items included in this list.

In the "A" category, 5–10% of the products account for 70–75% of the overall material consumption.

The "B" category includes the remaining 15–20% of the goods, which account for 15–20% of the overall material consumption.

Five to ten percent of materials will fall into the "C" category. Hence, the remaining products are considered surplus.

Based on their relative positions, we may deduce that objects in Category A need the tightest supervision, items in Category B can get by with less care, and item C can be left relatively unmonitored [23].

4   Analysis and Results

Power BI enables the prediction of the sales of 2021 in the appliance manufacturer using historical data of their sales during the previous two years. As shown in Fig. 1, these are the sales of all the items shown in the slicer, and by using the slicer, they can click on each item individually or together to show their predicted sales. The pattern in Fig. 1 indicates that there will be a drop in sales during 2019. In January 2017, the sales were around 10.9 K; in January 2018, the sales were around 10.2 K. In January 2019, sales were predicted to drop to the lowest level, about 9.3 K. The highlighted area represents a confidence level of 95% means.

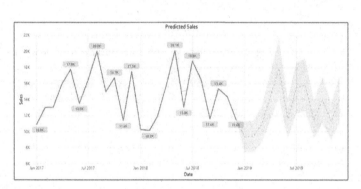

Fig. 1. Predicted sales using the exponential smoothing method

Additionally, Power BI has another time series method, linear regression. This model has been created by MAQ software. Forecast Using Multiple Models by MAQ Software enables the implementation of four forecasting models to learn from historical data and predict future values. The forecasting models include Linear Regression, ARIMA, Exponential Smoothing, and Neural Networks. This paper will only use linear regression from MAQ software because the other methods require more data points, which they could not provide. This visual is excellent for forecasting budgets, sales, demand, or inventory. The R package dependencies are forecast (offers methods and tools for displaying and analyzing univariate time series forecasts), plotly (makes interactive, publication-quality graphs online), zoo (key design goals are independence of a particular index, date, and time), lubridate (makes it easier to work with dates and times). As shown in Fig. 3, there will be a drop from the beginning of 2019. The previous method in Fig. 1 indicates the same situation in Fig. 2: a drop in sales at the beginning of 2019. The highlighted area represents a confidence level of 95% means.

Figure 3 depicts the same slicer as Figs. 1 and 2, but with what seems to be more information than in either of those earlier figures. The slicer is seen as a scatterplot, with

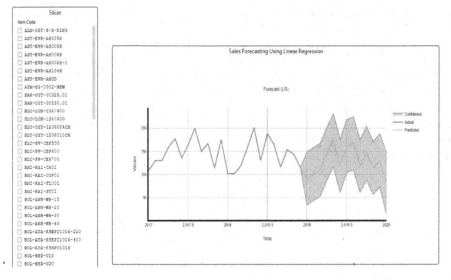

Fig. 2. Predicted sales using the linear regression method

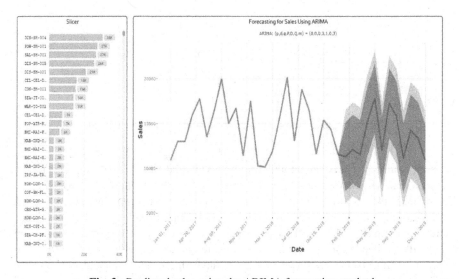

Fig. 3. Predicted sales using the ARIMA forecasting method

product IDs along the Y-axis and sales figures along the X-axis. For instance, DIS-SH-004, listed as the first item, has racked up sales of over $30,000 between 2017 and 2018. If you have a time series and want to apply a model to predict future values based on those values, you are engaging in time series forecasting. ARIMA is crucial for every business analyst when forecasting resources like budgets, sales quotas, marketing campaigns, and purchases. Better choices might be made with reliable projections. The current UI uses the widely used ARIMA technique of prediction. A broad study area is forecasting time series that can be made stationary using ARIMA models. ARIMA models attempt to capture the autocorrelations in the data, whereas exponential smoothing methods are based on a description of trend and seasonality in the data. In this context, "support" refers to the ability to use either a seasonal or a non-seasonal model. The settings of the algorithm, as well as the aesthetic characteristics, are within their control. The following R packages are required for the plot to appear: proto (enables a programming methodology called prototyping), zoo. Based on the Figure, it seems that the ARIMA method's sales projections for the beginning of 2019 will differ from those obtained by linear regression and exponential smoothing. Compared to January 2017 and January 2018 alone, however, January 2019 sales are expected to be higher. However, 2019 will see an overall decline in the data pattern. Colors denote different degrees of certainty; blue indicates 95% certainty, while grey represents 99% certainty.

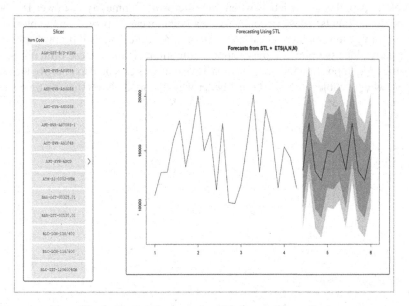

Fig. 4. Predicted sales using STL forecasting method

The STL was used to project 2019 sales for all products in Fig. 4. In addition, a slicer allows seeing the product's sales by clicking on a specific item. First, a non-seasonal forecasting algorithm is used for the seasonally adjusted data to generate forecasts of STL objects. Then, the seasonality using the last year of the seasonal component is re-calculated. STL offers several benefits compared to the traditional decomposition approach and the ARIMA technique. Unlike ARIMA, which only works with monthly and quarterly data, STL can deal with any seasonality. The user may choose how quickly the seasonal element evolves over time. The user may also adjust the trend-cycle smoothness to their liking. Since the user may choose a strong decomposition, it can withstand extreme values. Since this is the case, we may rest certain that our trend-cycle and seasonal components estimations will remain unaffected by rare odd observations. However, they will have an impact on the remaining. However, STL does not provide tools for subtractive decompositions and does not automatically manage a trading day or calendar variation. Colors denote different degrees of certainty; blue indicates 95% certainty, while grey represents 99% certainty. Figure. 4 predicts a sales decline between the third month and the end of 2019. While linear regression and exponential smoothing both anticipate a decline in sales throughout the forecast period, our technique suggests that sales will be relatively stable.

5 Findings

This paper implements four methods, exponential smoothing, ARIMA, linear regression, and STL, to determine the best forecasting technique. Concerning accuracy calculation, the STL method was excluded. This is because the actual sales could not be displayed in the plot. Also, the Y-axis labels where the Microsoft community of Power BI could not process the inputs. However, only six data points were displayed on the plot; only two were in the forecasting-shaded area. Additionally, the X-axis did not display the dates clearly since the plot can only show six durations. Despite these difficulties, the time-series decomposition feature in Power BI visuals was executed to improve the sales analysis, which enables visual data in a better way. Figure 5 shows that the time series decomposition model displays four plots by default.

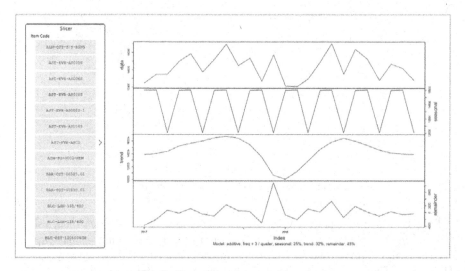

Fig. 5. Time series decomposition chart

The actual sales are shown in the first data plot. The second represents seasonality, which indicates the existence of fluctuations at regular intervals smaller than a year, such as weekly, monthly, or quarterly. Differences between cyclical and seasonal trends in a time series are discussed. The seasonal plot shows that the shifts happen every three months. Aside from that, it is stable for a while, then lowers, then increases, and the cycle repeats every three months. The trend, the rising or falling value in the series, is the third plot and represents the series' irregular component. This unique graphic also uncovered more data that is shown underneath the graphs. This is an additive. Therefore, the first chart is a summing of the other three. There is a 25% seasonal component, a 32% trend component, and a 43% irregular component, which is the error or residual value, and the frequency is every three months, which is quarterly.

Forecast inaccuracy on its own is not sufficient to rule out forecasting as a management tool since accuracy is the benchmark by which the optimal approach is determined. The precision, in hindsight, predictions of future values should be pretty close to the values that ultimately emerge. The benefits of squeezing an extra half a percent of accuracy out of forecasts that are already good enough to employ in decision-making may be minimal. Mean absolute percent errors (MAPE) and confidence intervals were generated to assess the closeness of the estimates to the firm's actual sales and the extent to which they reveal any unexpected trends. If sales projections are close to reality, then these numbers make sense. Additionally, Eqs. (1) and (2) provide the accuracy and bias formulae necessary to determine the accuracy of the forecast systems (2).

$$Accuracy = 1 - \left(\frac{ABS(error)}{Forecast} \right) \tag{1}$$

$$Bias = \frac{Actual}{Forecast} \tag{2}$$

Starting with the exponential smoothing, Table 1 shows the calculation of exponential smoothing's accuracy. The total sales forecast by the exponential smoothing model is 155,667 items, which means sales will drop by 11.21% with 86.83% accuracy.

Table 1. Accuracy of the exponential smoothing method

Period	Forecast	*Abs* (Error)	Bias	Accuracy
1	9350	1234	1.132	86.80%
2	9447	2135	1.226	77.41%
3	10855	1610	1.148	85.17%
4	14270	1621	1.114	88.64%
5	17924	1019	1.057	94.31%
6	11630	1648	1.142	85.83%
7	15479	2132	1.138	86.23%
8	15842	2400	1.151	84.85%
9	11507	1786	1.155	84.48%
10	13936	2101	1.151	84.92%

(*continued*)

Table 1. (*continued*)

Period	Forecast	*Abs* (Error)	Bias	Accuracy
11	11313	1560	1.138	86.21%
12	14114	414	1.029	97.07%
Total	**155667**	**19656**	**13.581**	**86.83%**

In Table 2, the calculations of the ARIMA model's accuracy show the total expected sales of 2019 are 162,695 items, which means a drop in sales by 7.20% with an accuracy of 92.16%.

Table 2. Accuracy of the ARIMA model

Period	Forecast	*Abs* (Error)	Bias	Accuracy
1	11317	1057	10.707	90.66%
2	12121	1971	6.150	83.74%
3	11543	357	32.333	96.91%
4	15225	571	26.664	96.25%
5	17813	2323	7.668	86.96%
6	11996	1035	11.590	91.37%
7	17253	1535	11.240	91.10%
8	15837	645	24.553	95.93%
9	11086	508	21.823	95.42%
10	14216	1144	12.427	91.95%
11	13388	957	13.990	92.85%
12	10900	651	16.743	94.03%
Total	**162695**	**12754**	**12.756**	**92.16%**

In Table 3, the calculations of the linear regression model's accuracy show the total expected sales of 2019 are 156,543 items, which means a drop in sales by 10.71% with an accuracy of 77.66%.

Table 3. Accuracy of the linear regression model

Period	Forecast	*Abs* (Error)	Bias	Accuracy
1	10099	161	62.727	98.41%
2	10982	832	13.200	92.42%

(*continued*)

Table 3. (*continued*)

Period	Forecast	Abs (Error)	Bias	Accuracy
3	14408	2508	5.745	82.59%
4	17460	1664	10.493	90.47%
5	11795	8341	1.414	29.28%
6	16128	3097	5.208	80.80%
7	16759	2029	8.260	87.89%
8	11810	4672	2.528	60.44%
9	14554	2960	4.917	79.66%
10	11390	3970	2.869	65.14%
11	13045	1300	10.035	90.03%
12	8113	3438	2.360	57.62%
Total	**156543**	**34972**	**4.476**	**77.66%**

Henceforth, from the information provided above in Tables 1, 2, and 3, a conclusion can be made that the Autoregressive Integrated Moving Average (ARIMA) model has the best accuracy among the other models.

Meanwhile, the error of each technique should be calculated using the MAPE formula as follows:

$$MAPE = Average(Error\%) \tag{3}$$

$$Error = \frac{ABS(error)}{Actual} \tag{4}$$

Tables 4, 5, 6, and 7 show the calculations of MAPE for all models.

Table 4. MAPE of exponential smoothing model

Period	Actual	Forecast	Abs (Error)	Error %
1	10584	9350	1234	11.66%
2	11582	9447	2135	18.43%
3	12465	10855	1610	12.91%
4	15891	14270	1621	10.20%
5	18943	17924	1019	5.38%
6	13278	11630	1648	12.41%
7	17611	15479	2132	12.10%
8	18242	15842	2400	13.16%

(*continued*)

Table 4. (*continued*)

Period	Actual	Forecast	*Abs* (Error)	Error %
9	13293	11507	1786	13.43%
10	16037	13936	2101	13.10%
11	12873	11313	1560	12.11%
12	14528	14114	414	2.85%

Table 5. MAPE of ARIMA model

Period	Actual	Forecast	*Abs* (Error)	Error %
1	10584	11317	733	6.93%
2	11582	12121	540	4.66%
3	12465	11543	922	7.39%
4	15891	15225	666	4.19%
5	18943	17813	1130	5.97%
6	13278	11996	1282	9.65%
7	17611	17253	358	2.03%
8	18242	15837	2405	13.18%
9	13293	11086	2207	16.60%
10	16037	14216	1821	11.35%
11	12873	13388	516	4.00%
12	14528	10900	3628	24.97%

Table 6. MAPE of the linear regression model

Period	Actual	Forecast	*Abs* (Error)	Error %
1	10584	10099	485	4.58%
2	11582	10982	600	5.18%
3	12465	14408	1944	15.59%
4	15891	17460	1570	9.88%
5	18943	11795	7148	37.73%
6	13278	16128	2851	21.47%
7	17611	16759	852	4.84%
8	18242	11810	6432	35.26%

(*continued*)

Table 6. (*continued*)

Period	Actual	Forecast	*Abs* (Error)	Error %
9	13293	14554	1262	9.49%
10	16037	11390	4647	28.98%
11	12873	13045	173	1.34%
12	14528	8113	6415	44.15%

As shown in those tables, the exponential smoothing method's MAPE is 11.48%, while the linear regression model has a greater MAPE of 18.21%, and the ARIMA model has the lowest MAPE of 9.24%. Thus, the recommended model for the company to use is the ARIMA as a forecasting technique that projects the future values of sales.

6 Discussion

Demand forecasting is an essential tool for businesses to predict the sales that decisions will be built upon. Also, it is important to have a well-managed inventory, meaning that overstocks and stockouts are the major problems that could occur if demand forecasting is not applied. Demand forecasting has two main types, which are Quantitative forecasting and Qualitative forecasting. This paper used four quantitative forecasting techniques, Linear Regression, Exponential smoothing, ARIMA, and STL, where historical data was used to build future plans using statistical methods to have the most accurate outcomes.

It was determined which of the four approaches was more accurate by comparing their error rates using Mean Absolute Percent Error (MAPE) and the accuracy of their respective forecasts. Based on the findings, it is clear that ARIMA is the most reliable method for predicting a series' future values using just the series' momentum. If you have at least 40 historical data points showing a steady or continuous trend over time with few outliers, you'll get the best results using this method for short-term forecasting. ARIMA, often known as Box-Jenkins after its original inventors, outperforms exponential smoothing approaches when data is sufficiently lengthy and the correlation between prior observations is stable. The ARIMA approach produced more accurate and reliable findings compared to the other methods. Using the ARIMA model, the margin of error was 9.24%, while the precision was 92.16%. On the other hand, specific strategies may do better than others when dealing with brief or highly volatile data.

7 Limitations

Implementing a forecasting method may not be doable for some businesses that lack retrieving appropriate sales data for previous years, which is an essential element of forecasting to recognize patterns. Also, it would be better for researchers to perform forecasting using multiple software programs where some features and model processing limitations may occur.

8 Conclusions

In conclusion, this study aimed to evaluate different forecasting techniques for demand projections in the food equipment industry. Accurate forecasting is critical for businesses to manage inventory levels and effectively meet consumer demand. By analyzing historical sales data using time series methods, the ARIMA approach demonstrated the highest forecasting accuracy compared to other commonly used techniques. The findings carry important implications for inventory management and supply chain planning practices. With more reliable sales projections, businesses can better allocate resources to avoid stockouts or overstocking. This helps reduce costs while still satisfying customer needs. The ARIMA method provides a data-driven tool to support such strategic decision-making.

Some opportunities for further research could involve expanding the analysis period to incorporate additional years of sales data. This may lead to even more precise forecasts, especially if accounting for longer-term trends. Additionally, future work could explore combining qualitative approaches with quantitative methods. Incorporating salesperson inputs or consumer surveys may help capture hard-to-predict market changes. Finally, applying these forecasting models to other sectors beyond food equipment could reveal different industry demand patterns. Overall, this study contributes to the need for predictive demand modeling, especially as businesses aim to operate more efficiently through data-driven operations management. The promising results of ARIMA forecasting suggest its suitability for supply chain planning within food equipment and related industries.

References

1. Wedowati, E.R., Singgih, M.L., Gunarta, I.K.: Production System In Food Industry: A Literature Study (2014)
2. Wedowati, E.R., Singgih, M.L., Gunarta, I.K.: Determination of modules in pleasurable design to fulfil customer requirements and provide a customized product in the food industry. Designs 4(1), 7 (2020)
3. Archer, B.: Demand forecasting and estimation. Demand Forecast. Estimation. 77–85 (1987)
4. Kurzak, L.: Importance of forecasting in enterprise management. Adv. Logistic Syst. 6(1), 173–182 (2012)
5. Penttilä, T.: Demand forecast process as a part of inventory management. Turku Univ. Appl. Sci., 45 (2009)
6. Putra, L.: Qualitative Forecasting Methods and Techniques. Retrieved October 20, 2018, from Accounting Financial & Tax website (2012). https://bit.ly/2RIX4IE
7. Mondal, P., Shit, L., Goswami, S.: Study of effectiveness of time series modeling (ARIMA) in forecasting stock prices. Int. J. Comput. Sci. Eng. Appl. 4(2), 13 (2014)
8. Dargar et al.: Applied ARIMA and optimized its parameters using genetic algorithm to forecast demand for auto parts. Their ARIMA-GA model outperformed other approaches. This recent study supports the use of ARIMA (2022)
9. Tseng et al.: Combined ARIMA with LSTM neural networks for stock price forecasting. Their hybrid model achieved higher accuracy than single techniques. This points to potential value in combining methods (2021)

10. Çelik et al. Applied STL, ARIMA and SVM to forecast steel industry demand in Turkey. STL-SVM provided the best fit, highlighting STL's suitability for industrial forecasting (2020)

11. Zhai et al.: Integrated ARIMA with deep learning LSTM models for electric vehicle charging demand prediction. The ensemble approach outperformed individual models (2020)

12. Livieris et al.: Combined linear regression, SVR and RNN for olive oil production forecasting in Greece. RNN exhibited the best performance. This shows interest in hybrid artificial intelligence techniques (2019)

13. Taylor, J.W., Snyder, R.D.: Forecasting intraday time series with multiple seasonal cycles using parsimonious seasonal exponential smoothing. Omega **40**(6), 748–757 (2012)

14. Indrasen, Y., Rajput, V., Chaware, K.: ABC Analysis. Adv. Res. Appl. Sci. **5**(5), 134–137 (2018)

15. Qin, L., Li, W., Li, S.: Effective passenger flow forecasting using STL and ESN based on two improvement strategies. Neurocomputing. **356**, 244–256 (2019)

16. Samarawickrama, I.: Electricity demand prediction of large commercial buildings using support vector machine. Retrieved (8 May 2019) (2014). https://bit.ly/2LBXGyy

17. Jury of executive opinion. (nd.) *Farlex Financial Dictionary*. Retrieved January 5 2019 from The Free Dictionary website (2009). https://bit.ly/2XHDKAM

18. Business J.: *What Is Market Survey?* Retrieved January 19, 2019, Business Jargons website (2016). https://bit.ly/2XcHHy2

19. Zigu.: *Sales Force Composite Definition*. Retrieved January 19, 2019, MBA Skool-Study.Learn.Share. website (2011). https://bit.ly/30339mh

20. Twin, A.: *Delphi Method*. Retrieved January 19, 2019, from Investopedia website (2019). https://bit.ly/2wgaBh0

21. Bartleby Research. *Associative and Time Series Forecasting Models*. Retrieved February 2, 2019, from Bartleby.com website (2009). https://bit.ly/2xuK0xn

22. CFI. *Forecasting Methods - Top 4 Types, Overview, Examples*. Retrieved February 2, 2019, from Corporate Finance Institute website (2016). https://bit.ly/2JjD77g

Simulation Model for the Optimization of Preparation Times in a Metalworking Industry Using Single Minute Exchange of Dies

Dalma J. Valverde Alania[⊠], Jakelin B. Cuellar Gonzales,
and Rubén D. Arzapalo Bello

Continental University, Huancayo 12000, Perú
{71784357,72554067,rarzapalo}@continental.edu.pe

Abstract. This article addresses the problem of matrix changes in a metalworking industry; due to the requirement to fulfill the customer's production orders on time, planning is unsustainable, thus causing delays in deliveries, low production, too much waste and loss of credibility with the customer. In the production process of kitchens, the sausage process is the one that generates more time, which leads to higher cost overruns and a lower productivity index, the methodologies applied as SMED "Single Minute Exchange of Die" are efficient, but they lack to adapt them to a multiformat change process, that is why a programming model was designed using the FlexSim software taking as a starting model the process with the initial times and obtaining a final one where it was possible to minimize costs in the same way a greater control of stock and an increase in the production of kitchens due to the reduction of preparation times was obtained; the study was validated in a kitchen factory where the kitchen with the highest demand was chosen, obtaining results with a reduction of 48.5% of the Setup, making it more productive times and a growth in the production of kitchens of 33.8% could be evidenced.

Keywords: SMED · preparation time · optimization · production

1 Introduction

Nowadays, in manufacturing, a current problem is machine stoppages due mostly to rework, format changes that delay the production process, resulting in products not being delivered on time.

In Peru, it is mentioned that the growth of industrial companies amounted to a figure of more than 14.3% in the last year, due to the technification of their production processes within the metalworking sector, however, the adaptation to the new process generates problems and delays production [1]. The main difficulty observed is the lack of planning and abruptly changing the format, causing them to make wrong decisions or solve a problem only for the moment, but not be able to eliminate it [2]. To do this, there are different methodologies or tools such as SMED, which is beneficial to be able to find activities.

© The Author(s), under exclusive license to Springer Nature Switzerland AG 2024
S.-H. Sheu (Ed.): IEIM 2024, CCIS 2070, pp. 30–42, 2024.
https://doi.org/10.1007/978-3-031-56373-7_3

The SMED methodology is an important technique of continuous improvement that explains the development of the machine readiness study. "SMED is an approach or philosophy that generates rapid changes, it means changing the matrix in less than 10 min" [3]. This tool was created by Shigeo Shingo at the Toyota company, it helped to decrease machine set-up times to a considerable time. The application of the tool is important because if it is implemented correctly and all activities that delay the process are eliminated, they would not only be increasing the availability of the machine but also the production efficiency, which generates greater profits.

It is mentioned that the SMED tool is beneficial because it allows you to have a clear vision, therefore, applying this tool provides and gives solutions to eliminate waste or help find activities that generate problems within the company. One of the most commonly used methods is a Pareto chart, which results in the tool eliminating waste and helping to identify problematic activities within the company. It is recommended to use a Pareto to identify the main problems within the company [2]. The use of tools such as causal diagrams, Pareto and formats for activities, a great improvement in productivity is obtained by 32%, this is due to the fact that it is possible to reduce the minutes in the machine studied, this was achieved thanks to a thorough study to understand the process times, improved efficiency by 20%, Since by decreasing the calibration of the machine the empty time is less, the methodology helped to reduce downtime and increased efficiency in the machines.

This research aims to reduce the preparation times of a hydraulic press in a kitchen manufacturing company through the use of the SMED methodology, thus increasing productivity and therefore increasing the profitability of the company. For this case study, the FLEXSIM simulation program will be used where the production process of kitchen manufacturing will be designed, this simulation environment will help analyze the processing times of each activity, providing an overview to detect possible deficiencies in the process and apply improvements. The simulation in FLEXSIM will allow us to propose different scenarios that will contribute to decision-making within the production process.

The research work is structured as follows: as a beginning is the introduction to the research topic with the objective of the study. Section 2 presents the literature review, Sect. 3 presents the case study, Sect. 4 mentions the development of the proposal. Section 5 presents the evaluation of results, compares the processing times before and after applying the SMED and finally the conclusions and recommendations of the research paper.

2   Literature Review

For this point, information will be collected from experts in the area and starting as a first point, it is mentioned that the great demand that consumers have today will determine the production model, for this it is important to implement the SMED methodology, allowing to significantly reduce the time in preparation operations. The SMED methodology can bring a lot of superiority to a company, such as improving quality, production flexibility, reducing downtime, production size and movements [4].

The application of the FlexSim to simulate the events within the development of production will allow the evaluation of its production system in a purely virtual environment, with the aim of achieving maximum efficiency within the company. This process requires the generation of an artificial history of the production process with the help of observation in order to have a relationship with the real characteristics, allowing the best decisions to be made within the production process [5].

The SMED methodology allows the implementation of a continuous improvement action plan using the tools of lean manufacturing, with the sole purpose of eliminating wasted time, by applying this methodology it can be noticed in advanced and developing countries, significant improvements in operational efficiency are achieved, thus achieving a competitive advantage in general.

On the other hand, we have research studies that were applied where it was aimed at determining alternatives for improvement in the process of changing tools in the lines of picks and axes, methods such as: filming, observation sheets, analytical cursogram, surveys were used where it was concluded that the SMED managed to reduce the process time and also improved the optimization of resources [6].

In the same way, it was used for the change of matrices following the stages established to perfect the activities that are developed in the machine, for the study the following methods were applied: 5S, observation sheets and check list from which it was concluded that the SMED achieved a reduction of 66.29% and an improvement in production capacity of 28.91% on average [7].

Likewise, in a study carried out in an automotive industry, the steps to be carried out for a correct application of the SMED methodology were detailed using methods such as: Diagram of process activities and a Gantt chart to compare time changes, where it was concluded that the SMED achieved an optimal time reduction of 0.08 min since before the time lasted 0.29 and now it is 0.20 there is an improvement of 21% [7].

It is essential that all organizations know what their main losses are and thus identify the best way to eliminate them. SMED allows us to carry out measurements in the production process, which allows us to identify its maximum losses. The study carried out will allow the identification and combat of losses with the team involved. Considering that all processes are identified by the flow of materials and operations, it is necessary to establish a pleasant relationship between equipment and man [8].

Finally, a study points out that the present research aims to increase the productivity of UTC (Internal Container Terminal) using analysis tools through value stream mapping (VSM), SMED and 5S. Based on the VSM result, the highest waste is obtained in unnecessary movement 18.574% and document transport 18.154% with value-added activities of 60.81% and the percentage of non-value-added activities of 39.19% [9].

3   Case Study

The research focused on applying the steps of the SMED methodology "Single Minute Exchange of Dies" in a metalworking industry that is dedicated to the manufacture of kitchens, the company has two manufacturing lines: home line and industrial line where the product with the highest demand is the 2-burner table kitchen. Also, the company has 26 machines to supply all production, within the furniture area is the sausage process and there is the main machine that is the hydraulic press of 250 TN which will be the object of study since major problems were detected in this machine due to the constant changes of matrix that are made for the manufacturing process of the kitchens so it is It aims to reduce set-up times, increase production and make processes more agile. The SMED methodology was chosen because it provides many techniques to reduce matrix change times, change times are red times that is, times that are not paid by the client therefore they must be eliminated. Companies that want to maintain a competitive advantage must make processes more flexible and a fundamental process is to produce in small batches, but at strategic costs.

To carry out the study, an internal investigation was carried out extracting important data from the company and these were complemented with real data, the study focused on evaluating the matrix change times of the hydraulic press of 250 TN and to measure the production we chose to monitor the kitchen with the highest demand at the end we will see how the production of kitchens will be reflected after the SMED is applied.

For the application of the SMED methodology, each stage must be followed step by step together with the participation of all the collaborators of the organization, as shown in the image the preparation time before a new batch enters is what could be reduced by up to 90%, it is important to make a general follow-up (Fig. 1).

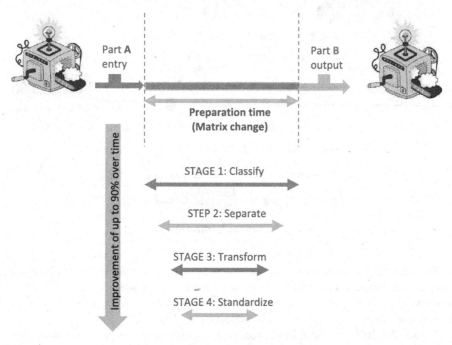

Fig.1. Variation of preparation times.

4   Development of the Case Study

The application of the SMED methodology will be carried out in a kitchen model of the home line that represents the kitchen with the highest demand for the company:

The two-burner table cooker is the one with the largest market share, order ranges from 350 to 450 stoves monthly.

As research instruments we will use the Check List of the equipment, record of control of quantity of pieces, chronometer for the analysis of times and the simulation model in FlexSim of the entire production process.

4.1   Identification of the Effect

According to the selected product, the delay in times to prepare the machine is identified. Next, in the image you can see the Ishikawa diagram tool that will help us to identify the effect in first, which is the delay in machine preparation times, to reach this result they had to detail their causative predecessors, because the proposed context focuses on the machine relationship, materials, method and labor (Fig. 2).

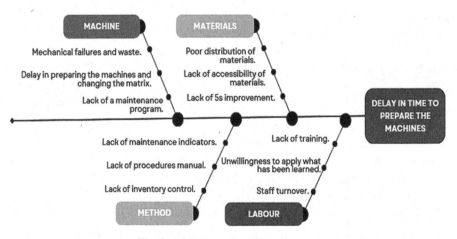

Fig. 2. Ishikawa cause-effect diagram.

To understand the magnitude of the problem, a graphic representation will be made using the continuous improvement tool called Pareto Diagram where the 80/20 rule will be shown [10]. In the representation of the Pareto diagram, it can be seen that 20% of the activities are the cause of 80% of the effects.

Within the 20% are the following activities: delay in changing dies, mechanical failures and waste; also, tools and materials are out of reach of personnel. The research study will focus on 20% of the aforementioned activities (Fig. 3).

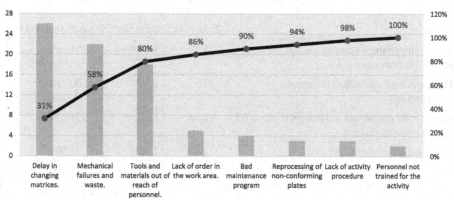

Fig. 3. Pareto chart of causative activities.

5   Evaluation of Results

According to the analysis raised above, the product chosen to study is the 2-burner table kitchen which goes through the sausage process where the hydraulic press is specifically located. In order to extract the information on the preparation times of the machine, a checklist was used to capture in detail the activities they perform in the course of the change from one die to another.

From which the following results were subtracted, (see Table 1) it is possible to observe the times it takes to carry out each activity and also the percentage participation that each one has in order to identify which of all the activities is the one that presents the longest time to be carried out in the sausage process.

Table 1. List of activities

Activities	Time (min)	Percentage %	Accumulated %
Communicate the matrix change of the next product	5.6	5.00%	5.00%
Request permissions to shut down the machine	10.05	8.98%	13.98%
Move to the sausage area	3.67	3.28%	17.26%
Verify that the machine is not energized and that the components are in good condition	5.81	5.19%	22.45%
Move to the office to enlist and fill out the documentation for the next piece	12.54	11.20%	33.65%
Move to the hydraulic press	3.2	2.86%	36.50%
Prepare the tools to perform the array change	5.5	4.91%	41.42%
Remove the current die from the hydraulic press	5.9	5.27%	46.69%

(continued)

Table 1. (*continued*)

Activities	Time (min)	Percentage %	Accumulated %
Search among the 33 dies and select the matrix for the 2-burner table kitchen	5.23	4.67%	51.36%
Place, position, and adjust the matrix (reformatting)	10.5	9.38%	60.74%
Move to the warehouse and bring the plates for the next type of product	12.56	11.22%	71.95%
Clean and lubricate plates	4.53	4.05%	76.00%
Clean the base of the hydraulic press where the iron will be placed	2.51	2.24%	78.24%
Lift, settle and place the plate in the hydraulic press	10.51	9.39%	87.63%
Request permissions to unlock the machine	12.82	11.45%	99.08%
Start the machine so that it outputs the first piece of the second product	1.03	0.92%	100.00%
TOTAL	111.96	100%	

5.1 Step 1: Separate Internal and External Activities

To achieve effectiveness in the first stage, it is recommended to order the activities following the flow of the process then use the research tools to identify and order the activities that will be selected into two groups:

Internal: These are the activities that are carried out with the machine stopped.

External: These are the activities that are carried out with the machine in operation.

Each activity will be evaluated in order to considerably reduce the time, it should be noted that not all internal activity becomes external there are operations that must necessarily be performed with the machine off, there are also operations that no matter how long it takes to perform them, they must be done because the working mode is already established, take into account the assembly and disassembly of the pieces.

Of the 16 activities, 3 were classified as external (see Table 2) with a time of 16.68 min.

And the rest of the activities as internal with a total time of 95.28 min (see Table 3) the list of activities that will be analyzed and then evaluated if it remains as internal or becomes external is observed.

Table 2. Separation of External Activities

Activities	Time (Min)	Guy
Communicate the matrix change of the next product. (D)	5.6	External
Request permissions to shut down the machine. (D)	10.05	External
Start the machine so that it outputs the first piece of the second product. (D)	1.03	External
TOTAL	16.68	

Table 3. Separation of Internal Activities

Activities	Time (Min)	Guy
Move to the sausage area. (T)	3.67	Internal
Verify that the machine is not energized and that the components are in good condition. (I)	5.81	Internal
Move to the office to enlist and fill out the documentation for the next piece. (T)	12.54	Internal
Move to the hydraulic press. (T)	3.2	Internal
Prepare the tools to perform the array change. (O)	5.5	Internal
Remove the current matrix from the hydraulic press. (O)	5.9	Internal
Search among the 33 dies and select the matrix for the 2-burner table. (O)	5.23	Internal
Place, position, and adjust the matrix (reformatting). (O)	10.5	Internal
Move to the warehouse and bring the plates for the next type of product. (T)	12.56	Internal
Clean and lubricate the plates. (O)	4.53	Internal
Clean the base of the hydraulic press where the iron will be placed. (O)	2.51	Internal
Lift, settle and place the plate in the hydraulic press. (O)	10.51	Internal
Request permissions to unlock the machine. (D)	12.82	Internal
TOTAL	95.28	

5.2 Step 2: Convert Internal to External Activities

The 13 activities that were classified as internal began to be evaluated individually together with the company's staff after carrying out the analysis, improvements were identified, and action plans were programmed where the processing times of each activity were optimized.

This involved making a study of times and movements to improve the transfer times of materials, automate the process of cleaning, loading and positioning of the dies, improve the procedure of intervention of energies and reinforce the practice of 5S.

After the improvements implemented and conversion of internal to external activities (Table 4), the following results were given:

Table 4. Extract of activities converted from internal to external.

Figure	Type of activity	Quantity	T. Operative (Min)
	Operations	7	33
	Inspections	1	5
	Transport	4	20.8
	Delays	4	26.68
	TOTAL	16	85.48

Before the application of SMED the processing time was 111.96 min.
After the application of the SMED the processing time would be 85.48 min.

5.3 Step 3: Improve Internal and External Activities

This stage will help to find the activities with times that we consider may become unnecessary, if we evaluate the external activity (see Table 5) we can see that it takes 5.6 min to only communicate the change of a matrix, what is proposed is to make a programmed plan for the change of matrix and implement communication radios so that the operator can give notice to the production area about the change of matrix from his point of work and not be mobilizing in person to the production area.

Table 5. Separation of External Activities.

Activities	Time (Min)	Guy
Communicate the matrix change of the next product. (D)	5.6	External

If we apply this improvement to reduce communication time and unify it with the improvements made (see Table 4) the results would be favorably affected (Table 6), obtaining the following:

With the change from internal to external activities, a time of 85.48min was obtained.

And with this last improvement applied it would be possible to reduce to 81.88 min.

With this result it is possible to determine the reduction of times that would be obtained after implementing the SMED, the preparation time of the hydraulic press took a start of 111.96 min.

With the application of the SMED the preparation time would be 81.88 min.

This indicates that a reduction of 26.87% would be achieved in the time it takes to change a matrix in the hydraulic press that in minutes would be 30.08 min.

Table 6. Extract of activities converted from internal to external.

Figure	Type of activity	Quantity	T. Operative (Min)
	Operations	7	33
	Inspections	1	5
	Transport	4	20.8
	Delays	4	23.08
	TOTAL	16	81.88

5.4   Processing Times and Production Capacity

It has as information that each 2-burner table cooker requires 12 pieces to be manufactured and that the daily production capacity of the hydraulic press is 500 pieces per day. Considering that 10 matrix changes are made per month for the manufacture of the kitchen.

Processing and capacity data: Before the application of the SMED monthly 730 kitchens were produced working with a SETUP time of 18.66 h / month with a machine production capacity of 12000 pieces per month. After the application of the SMED it would be possible to reduce the SETUP time to 13.64 h / month considering 10 changes of matrix / month with a working schedule of 192 h / month which results:

$$HydraulicPressProductionCapacity = \frac{12000(178.36)}{173.34}$$

$$HydraulicPressProductionCapacity = 12347 pieces.month \qquad (1)$$

Resulting in an increase in production capacity of 12347 pieces / month considering that 12 pieces enter for the manufacture of a 2-burner table kitchen, 1028 stoves per month would be manufactured.

5.5   Modeling of the Production Process in FlexSim Software

Production after deploying SMED.

The output of finished products was simulated in the Flexsim program, obtaining 1028 kitchens because of the previous evaluation (Fig. 4).

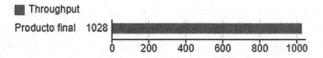

Fig. 4. Finished product indicator.

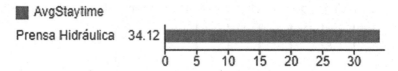

Fig. 5. Hydraulic Press Processing Time Indicator.

Considering the SETUP time of 81.88 min with 10 changes per month which leads to a daily SETUP time in a workday of 34.12 min (Fig. 5).

Likewise, the processing time required for changing dies in the hydraulic press was simulated in the Flexsim program. According to the evaluation, it takes 34.12 min to change the die (Fig. 6).

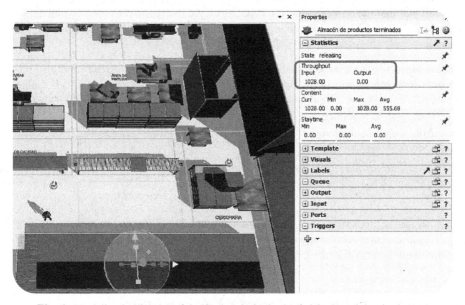

Fig. 6. Modeling in Flexsim of the final quantity in the finished product warehouse.

6   Conclusions and Recommendations

With the simulation in FlexSim and the application of the SMED methodology, the company will be able to reduce the preparation times of the machine by 26.87%, which initially had 111.96 min for the sausage process now with the application of the methodology would be reduced to 81.88 min. In terms of production, an increase in the capacity of the hydraulic press would be achieved by 12347 pieces / month, reaching 1028 kitchens per month. The implementation of the SMED methodology throughout the company would provide great contributions (Fig. 7), such as:

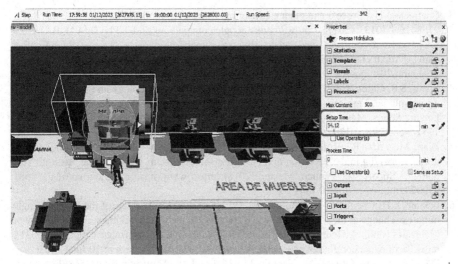

Fig. 7. Modeling in Flexsim of the processing time of the hydraulic press.

- Better responsiveness in customer service, as the requirement can be covered with shorter cycles in manufacturing and order delivery.
- Decrease in machine preparation times and die change. Also, idle time in equipment and machines.
- Greater use of materials without having a lot of waste.
- Improvement in their productivity indicators, by increasing the output and reducing downtime in each preparation of machines.

The simulation with the FlexSim software allows us to identify in two different scenarios how the process is carried out with and without SMED, which implies that the use of technology helps us to have a forecast of how important this implementation. The most important thing about SMED is that you can combine the different techniques of continuous improvement, in order to make companies have better results and be competitive in the labor market.

Nowadays having an advantage at the level of process improvement are very important points for the growth of a company, more in the case of emerging companies in the metalworking sector of the Mantaro Valley, where each improvement can be projected for future investment projects.

References

1. INEI 2016. PERU: Business Structure (2016) https://www.inei.gob.pe/media/MenuRecur sivo/publicaciones_digitales/Est/Lib1445/libro.pdf. Accessed 5 May 2020
2. Aviles, N.M. Caiza, M.K. Reduction of waiting times in the blowing area of plastic containers by applying the SMED methodology. Postgraduate thesis. Milagro State University, Milagro, Ecuador (2019) http://repositorio.unemi.edu.ec/handle/123456789/4787. Accessed 10 May 2020

3. Cuatrecasas, L.: Process and plant engineering. Barcelona: profit Editorial I., S.L., pp. 3–71. 9788416904013 (2017). https://books.google.com.pe/books?id=CPNyDgAAQBAJ&pg=SA3-PA71&dq=que+es+smed+segun+shigueo+shingo&hl=qu&sa=X&ved=0ahUKEwiE58uU__XpAhVDK7kGHdewCAMQ6AEIODAC#v=onepage&q=que%20es%20smed%20segun%20shigueo%20shingo&f=false. Accessed 20 May 2020

4. Wang, S.S, Chiou, C.C., Luong, H.T.: Applying SMED methodology and programming in a low-volume, high-mix production model to reduce setup time: an s company case. IOP Lecture Series. Materials Science and Engineering, **598**(1), 012058 https://doi.org/10.1088/1757-899x/598/1/012058. Accessed: 10 May 2023

5. Diaz, M, Cruz, R., Román, R.: FlexSim simulation, a new alternative for engineering decision-making in the operation of a multi-test station system. online article. instituto Tecnológico Superior de Pánuco, Mexico (2018) https://www.redalyc.org/journal/614/61458109002/html/. Accessed 10 June 2023

6. Gómez, L.F., Ospina, D.Y., Ovalle, A.M., Villanueva, T.A.: Analysis of the SMED (single minute exchange of die) methodology in a company in the metalworking sector in the city of Manizales [online]. Online article. Universidad Autónoma de Manizales, Colombia (2021). https://acofipapers.org/index.php/eiei/article/view/1897/1616. Accessed: 10 June 2023.

7. Domínguez, A.B., Ortiz, D.M., Naranjo, I.E., Llugsa, J.M.: Application of the SMED methodology in the process of changing matrices in the metalworking industry: case of Ecuador [online]. Online article. Technical University of Ambato, Ecuador (2020) https://www.proquest.com/docview/2472669151. Accessed 20 July 2023

8. Parisotto, C., DE, J., D. SMED method: analysis and improvement [online]. Online article. Ritter dos Reis University Center - UniRitter, Porto Alegre, Brazil (2016) https://www.researchgate.net/publication/316221281_SMED_method_Analysis_and_improvement. Accessed 20 July 2023

9. Kusrini, E., Noyvanti, A.: Improving Productivity for Unit Terminal Containers Using Lean Supply Chain Management and Single Minute Exchange of Dies (SMED): A Case Study at the Port of Semarang in Indonesia. Online article. Islamic University of Indonesia (2020) https://publisher.uthm.edu.my/ojs/index.php/ijie/article/view/3745. Accessed: 18 June 2023.

10. Sanyapong, D.W, Coit, B., Anupong, A.K,: A review of Pareto pruning methods for multi-objective optimization, Computers & Industrial Engineering, vol. 167, 108022, ISSN 0360–8352 (2022). https://doi.org/10.1016/j.cie.2022.108022 (https://www.sciencedirect.com/science/article/pii/S0360835222000924)1997

Improvement Model to Optimize Packing Times in a Peruvian SME Agricultural Export Company Using Cellular Manufacturing, SMED and Standard Work

Edward Ocampo-Leyva[iD], Rafael Lopez-Morales[iD], and Alberto Flores-Perez[(⊠)] [iD]

Facultad de Ingeniería, Universidad de Lima, Lima, Peru
{20193173,20182830}@aloe.ulima.edu.pe, alflores@ulima.edu.pe

Abstract. The agricultural export business in Peru has grown considerably in recent years, playing a key role in the development of the nation's economy. To adapt to a growing market, companies must optimize their processes not to lose competitive edge. Problems such as excessive production times, low efficiency, and loss of the fruit due to rot are frequently met, these liabilities originate from, among some other causes, a high cycle time for the packing process, which can lead to negative economic impact and a worsening of the company's image. In this research, an improvement model using Lean manufacturing tools namely Cellular Manufacturing, SMED and Standard Work is proposed. The objective of the model is to optimize packing times, seeking to reduce the duration of the time needed to dispatch a crate of mangoes to at least 10.20 min, which following the implementation of the improvement model obtained a value of 9.84 min. Additional improvements were achieved such as a 2.19-min reduction for setup times, 11.9% reduction in the setup time ratio, and 8.9% reduction in moving times. The results obtained, validated by a simulation run in a controlled environment in Python 3.10, proved the effectiveness of the proposed model.

Keywords: Cellular Manufacturing · SMED · Standard Work · Packing Time · Python

1 Introduction

The agricultural export business has shown a considerable growth in recent years; in Peru, it went from representing 2.1% of the Peruvian GDP in 2010 to 4% in 2022 [1]. One of the agricultural goods Peru produces is mango (Mangifera indica), which exportations exceeded 240 000 tons in 2022 and its productive chain registered a FOB value of USD approximately 324 million in the same year. Allegedly, 88% of the producers of the mango exporting chain belonged to family agriculture, numbers that generate great expectations for the small producers, because of the rentability and growth opportunity [2]. This activity, therefore, will be generating more employment opportunities in the coming years, which may help to fight the increasing poverty that has reached 25.7% of Peruvian population in recent times [3].

© The Author(s), under exclusive license to Springer Nature Switzerland AG 2024
S.-H. Sheu (Ed.): IEIM 2024, CCIS 2070, pp. 43–53, 2024.
https://doi.org/10.1007/978-3-031-56373-7_4

Within the mango exporting chain, packing is one of the main activities, and sector companies must be aware that its optimization is key to remain competitive in a growing market. In the context of this research, the problem was related to excessive times of packing in a Peruvian SME, which lost a considerable amount of its employees due to budget cuts caused by the unplanned stoppage of activities during the Covid-19 pandemic. In 2022, the packing area recovered employees to a close number to what it had in previous years, with the objective of increasing production, however, registered times shown a 42.65% increase in the average cycle time compared to its pre-pandemic values. In literature research, it was found that it is possible that high cycle times in the sector can be caused by lack of resources, accumulation of tasks, unbalanced work lines and uncontrolled processes. This may result in higher cycle times for packing, least efficiency in the process, and wastes, which only reinforces the need for an improvement [4]. Additionally, it is known that the perishability of fresh products justifies seeking any improvement in packing, as long as it shows some benefit and can be economically justified [5]. For the reasons exposed, the company will have to better its packing time to prevent negative impacts. Therefore, to achieve this goal, a model was designed considering the implementation of Lean Manufacturing tools. First, Cell Manufacturing was used to improve the workflow by grouping activities in autonomous cells. Consequently, SMED tool was considered to reduce setup times in the new workflow. Finally, Standard Work tool helped to establish protocols, document, monitor the correct application, and ensure to maintain the improvements in the long term. The model was constructed drawing on similar studies that utilized the mentioned tools in comparable sectors, given the scarcity of information within the sector of its own.

Most of scientific papers examined do not provide precise information concerning the issue of packing time in the agricultural export industry, especially in the case of mango. This lack of clarity encourages us to make this research, aimed at proposing an enhancement. The research is organized into the following sections: Introduction, state of art, contribution, validation, discussion, and conclusions.

2　State of Art

2.1　Cellular Manufacturing

Cellular Manufacturing is a production strategy that organizes operations in autonomous work cells, where machines and operators combine efficiently to produce specific products [6]. This approach seeks to improve flexibility, reduce production cycle times, and minimize inventories, thus optimizing efficiency and quality in the manufacturing process. This can be evidenced in a research article which performed the tool on the production line of standard parts, where they improve efficiency by achieving a 31.3% reduction in the time for an order to be ready with their best variant model in comparison to a classic production-to-order scenario [7].

2.2　SMED

SMED, acronym for Single Minute Exchange of Die, is a tool based on the reduction of setup times in a production environment. Its main goal is to improve the efficiency and flexibility of production processes by optimization of setup activities, which if remain uncontrolled, may originate unproductive times and movements. This tool proved effective in different sectors, such as Manufacturing, where a study taken place in shoe mold sector achieved a reduction of 60% in the total setup time [8], whilst another study performed in the same sector achieved 37.8% [9]. Results may vary depending on the number and complexity of the setup activities.

2.3 Standard Work

Standard Work is a methodology that establishes clear and defined standards to carry out tasks efficiently, thus eliminating superfluous movements and optimizing workflow [10]. The implementation of Standard Work not only seeks to optimize efficiency, but also to improve and facilitate the training of new employees, reduce errors, and enable consistent performance measurement. In addition, this tool serves as a basis for continuous improvement, as it provides a clear reference for identifying opportunities for optimization and adjustment [11].

3 Contribution

3.1 Model Basis

In the present-day context, with the implications of the dynamics of a growing market, it's crucial for agricultural exporting companies to be able to satisfy market demands and expectations on time. The model proposed, which is designed to address the issues that lead to high packing cycle times, is built upon a thorough review of existing literature to find the most effective tools. The tools found to be most applicable to our research context include Cellular Manufacturing for leveling the working lines, optimizing the workflow, and reducing the cycle time, SMED for improving added times related to setup activities, and Standard Work for developing efficient protocols, establishing standards, and ensuring the sustainability of the improvements in the long term by training the implicated workers within a continuous improvement work frame. A matrix comparing the authors and the tools used to build the model is provided in Table 1.

Table 1. Comparison table of the proposed components and the state of art.

Articles/Components	Work lines leveling	Reduction of setup time	Movements Reduction
Rewers, P., Diakun, J. (2021)	Cellular Manufacturing		
Bocanegra-Herrera, C., Orejuela-Cabrera, J. (2017)	Cellular Manufacturing		

(*continued*)

Table 1. (*continued*)

Articles/Components	Work lines leveling	Reduction of setup time	Movements Reduction
Souza, J., Beluco, A., Biehl, L., Braz, J., Sporket, F., Rossini, E., Amaral, F., Ribeiro, R. (2019)		SMED	
Nedra, A., Néjib, S., Boubaker, J., Morched, C. (2022)			Standard Work
Improvement proposal	Cellular Manufacturing	SMED	Standard Work

3.2 Proposed Model

The proposed improvement model relies on three tools, which are Cellular Manufacturing, SMED and Standard Work. All three of these tools are intended to function in a cohesive and synergistic manner, as it is shown below in Fig. 1, which is the graphic representation of the model. However, special attention was given to Cellular Manufacturing as its direct objective is to reduce the packing cycle time, which in turn would directly impact the number of crates produced per day.

3.3 Components

Component 0: Diagnosis

To elaborate the improvement model, it was necessary to have a clear view of the problem and what was causing it. To achieve this, a diagnosis of the company was performed, finding that the main problem was the excessive times of packing. To validate it, a Value Stream Mapping was made for visualization purposes, as this tool gives a general view of the processes, its related time, and the sequence they follow [12], then a Pareto analysis helped to validate percentagewise the times of packing as the main issue, and finally, 5 Whys tool was implemented to find the problem's main causes. Additionally, a Problem Tree was also implemented.

Component 1: Cellular Manufacturing

Cellular Manufacturing was deemed as the first and main component because the tool impacts directly in the cycle time reduction. To implement this tool, four consecutive steps were taken. In the first step, the demand was measured, and the production rhythm was established. This production rhythm, also known as Takt Time, is the quotient of the available time for production and the previously measured demand during a determined period, in this case, a whole exporting campaign. Takt Time was determined to be 0.91 min per mango box. The second step consisted in reviewing the sequence of the processes in the packing line. Step three consisted in combining and leveling the processes, which from now on were grouped in four working cells using similarity and proximity criteria.

After that, calculations were made using Takt Time, standard times of the processes, and the number of available workers and machines to balance the working lines assigning the optimal number of resources to each cell. The fourth and final step was to design the blueprint for the cells. Cellular Manufacturing is expected to reduce the cycle time and improve the workflow by leveling the work lines [7].

Component 2: SMED

Even though the previously implemented tool was expected to make a significant improvement, it is heavily focused on the main productive cycle. However, other activities that add up to the total time of packing remain mostly untouched. That is why, as a second component of the improvement model, SMED was implemented to reduce unproductive times and waits mainly related to high setup times. To implement SMED, the first step consists in identifying setup activities. Then, the activities and its times are classified as internal or external in a matrix [13]. After that, activities must be observed and analyzed individually as the main goal is to reduce internal times by turning them into external, proposing an improvement for each of the ones that can be modified, even if this improvement requires additional resources like workers or equipment. It is important to notice that, during the analysis for externalization, it was discovered that few activities could be performed simultaneously, which represents an additional advantage in the matter of setup time reduction. At last, a numeric and percentual comparison is performed according to the activity, its classification, and the time it involves.

Component 3: Standard Work

Standard Work is the last component for this model, as it is considered as the optimal method to sustain the improvements of both Cellular Manufacturing and SMED on the long term. To implement it, it was necessary to document the sequence of the improved process and establish a protocol with it. This tool, which needs to be documented in a step-by-step manner, relies in the compromise to follow it, so it should be communicated and taught to all staff members through training. Adherence to this method should be monitored to ensure the sustainability of process' quality and efficiency achieved after the improvement model implementation. Standard Work will further aid in the execution of continuous improvement and effective problem resolution.

Component 4: Simulation

The last part of the model was aimed to validate the improvement. For this purpose, a simulation was run in a controlled environment in Python 3.10. This simulation considered the implementation of all three Lean manufacturing tools, giving a new packing cycle time and, consequently, a new number of crates produced per day.

3.4 Key Indicators

To evaluate the effectiveness of the improvement model, the following key indicators were selected.

- Packing cycle time: It shows the average time it takes to complete a cycle in the packing line, and thus dispatch a crate. The objective of the model is to reduce this

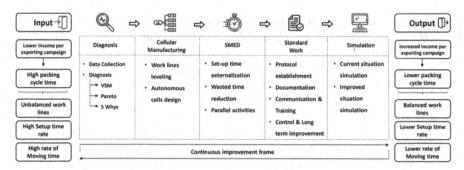

Fig. 1. Improvement model to reduce packing time

indicator to at least 10.20 min.

$$\text{Packing Cycle Time} = \sum \text{Process times for packing} \qquad (1)$$

- Setup time per cycle: Result of the sum of internal setup activity times required to complete a production cycle. Although its value is included in the packing cycle time, it is important to analyze it as an indicator of its own. The objective will be to reduce it in 50%.

$$\text{Setup time per cycle} = \sum \text{Internal setup activity times} \qquad (2)$$

- Setup time ratio: Shows the proportion of the setup time with respect of the cycle time. The objective is to lower it below 15%.

$$\text{Setup time ratio} = (\text{Setup time per cycle})/(\text{Packing cycle time}) \times 100 \qquad (3)$$

- Moving time ratio: Result from the proportion of the times spent moving and the total cycle time. The objective will be to reduce it to 20%.

$$\text{Moving time ratio} = (\text{Moving time})/(\text{Packing cycle time}) \times 100 \qquad (4)$$

4 Validation

4.1 Diagnosis

A packing cycle is not considered complete until a crate is dispatched from the packing line. Measurements made to the company show an average packing cycle time of 14.55 min, whilst the average value registered in years before pandemic was of 10.20 min. The gap, therefore, is deemed to be 4.35 min, which causes a yearly loss of 73597 USD, almost 7.9% of annual sales.

The main causes discovered were unbalanced work lines, high setup time rate, and high rate of moving time.

4.2 Validation Design

To design the validation, a macro review of the components was performed. First, data is collected, and a diagnosis is performed to identify problems and causes. Then, the first component is applied, which is Cellular Manufacturing, a production rhythm is established, then the processes are balanced and finally organized in work cells. The second component applied was SMED, for this, setup activities were classified as internal or external, improvements were made and some of the activities were turned into external. The next component is Standard Work, which focuses in maintaining the improvements made, establish a protocol and control further activities. Finally, a simulation was run in Python to obtain new values for the selected key indicators and compare them with their initial values.

4.3 Improvement Model Simulation

The simulation was run in Python 3.10, in a controlled environment, considering an 8-h work shift with 12 activities, including main processes and setups. The sequence of these activities was ensured to be maintained.

The parameters represented an improvement from the initial situation, which had 18 activities. The reduction of the number of activities was achieved by converting some internal setups to external ones and performing some others in parallel. The logic behind the simulation and the improved process sequence is shown below on Fig. 2.

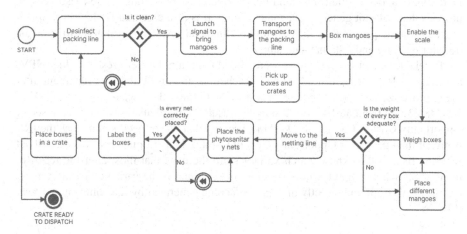

Fig. 2. Logic behind the simulation

The model's effect on the indicators is shown by comparing them before and after the implementation, as it is shown in Table 2.

It can be appreciated that the packing cycle time was reduced in 4.7 min, which represents a 32.4% decrease. There was a significant improvement in the setup time also, with a reduction of 2.2 min per cycle. The setup time ratio also decreased to 10.05%, which traduces in less time of the process spent in setups. All these key indicators accomplished

Table 2. Current situation vs Improved situation

Indicator	Before Improvement	After Improvement
Packing cycle time	14.55 min	9.84 min
Setup time per cycle	3.18 min	0.99 min
Setup time ratio	21.86%	10.05%
Moving time	30.93%	22.02%

the proposed objectives, the moving time ratio, although reduced, did not. However, it came close to the objective, and it still represented an important improvement. With the presented data it is observed that the improvement goals were met with the use of the simulators for the measurement of these indicators.

4.4 Economic Evaluation

Conducting an economic assessment is crucial to determining the viability of a project for several significant reasons. First, it provides a sound basis for informed decision-making, which is essential for investors, managers, and other stakeholders. It identifies opportunities and risks, avoiding erroneous investments and encouraging better decisions. In addition, an economic evaluation facilitates the efficient allocation of resources, including financial capital, time, and labor, which avoids wasting limited resources on projects that will not generate adequate returns. It is also essential for long-term planning, as it anticipates future cash flows and financial results, contributing to strategic decision-making and risk management.

For the financial validation, we were aimed to obtain the Net Present Value (NPV) and Internal Rate of Return (IRR) with the help of the profit and loss statement after the improvement and an economic cash flow. With these data, we obtained an NPV of 145498.22 USD and an IRR of 79% as the results. For calculations, we used a sector Opportunity Cost of Capital of 29% [14], proving that our project is viable, as the IRR is higher than the sector Opportunity Cost of Capital. In addition, a test was carried out with the risk tool, to know which variables are the ones that most alter the result of the test. Showing us that the dollars per kg of container is the variable that most alters the financial result and directly affects the income generated by the company per year (Fig. 3).

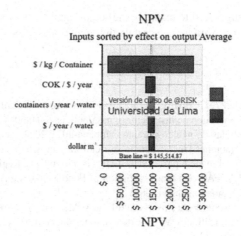

Fig. 3. Most impactful variables over the model viability

5 Discussion

The use of Cellular Manufacturing in a standardized product packing line could reduce cycle times by 31.3%, value highly consistent with the 32.4% obtained in the present research, also having leveled work lines as an integral part of the solution [7].

It is also mentioned that SMED can reduce 56.2% of internal times in a manufacturing company [15], in the present research, a reduction of 68.8% was achieved, with a couple less setup activities compared to those mentioned by the authors. Despite the differences, it can be noted in both cases that the improvement reduced internal setup times by more than half, which is a good indicator for SMED as an effective tool in different scenarios. It has also been said that standardization had a positive impact on process time reduction with the aid of handbooks and the periodical revision of the tasks performed by each operator [16], in other words, documenting and controlling as it is proposed in the current research.

6 Conclusions

The implementation and cohesive functioning of Cellular Manufacturing, SMED and Standard Work, facilitated to accomplish a cycle time reduction for mango packing to a value of 9.84 min, which represents a 32.4% decrease compared to the initial value of 14.55 min, and even surpassing the objective of reducing the indicator to 10.20 min by a tiny margin of 3.6%. As for other key indicators, setup time was reduced from 3.18 min to 0.99 thanks to the conversion of some activities from internal to external and the designation of parallel activities, the setup time ratio was reduced from 21.86% to 10.05%, which is also an important improvement because less time of the process is being invested in setups, also, the moving time ratio was reduced from 30.93% to 22.02%. The results show the significant improvement that the use of these tools had in an agricultural exporting company for its mango packing process.

Additionally, it is suggested to examine the sector again in the future, as the lack of information could be improved by newer data that may make results more precise.

References

1. Gob.pe.: MIDAGRI: Perú exportó más de 240 mil toneladas de mango durante campaña 2021–2022 (2022). https://www.gob.pe/institucion/senasa/noticias/604850-midagri-peru-exporto-mas-de-240-mil-toneladas-de-mango-durante-campana-2021-2022. Accessed 23 Nov 22
2. Organización de las Naciones Unidas para la Agricultura y la Alimentación (FAO): Manual para el mejoramiento del manejo poscosecha de frutas y hortalizas, Regional Office for Asia and the Pacific (RAPA) FAO and the Association of Food Marketing Agencies in Asia and the Pacific (AFMA), Santiago, Chile (1987)
3. World Bank: Rising Strong: Peru Poverty and Equity Assessment. https://www.worldbank.org/en/country/peru/publication/resurgir-fortalecidos-evaluacion-de-pobreza-y-equidad-en-el-peru. Accessed 23 Nov 23
4. Santosa, W., Sugarindra, M.: Implementation of lean manufacturing to reduce waste in production line with value stream mapping approach and Kaizen in division sanding upright piano, case study in: PT. In: X Analysis of Critical Control Points of Post-Harvest Diseases in the Material Flow of Nam Dok Mai Mango Exported to Japan, MATEC Web of Conferences, vol. 154 (2018). https://doi.org/10.1051/MATECCONF/201815401095
5. Yasunaga, E., Fukuda, S., Nagle, M., Spreer, W.: Effect of storage conditions on the postharvest quality changes of fresh mango fruits for export during transportation. Environ. Control Biol. **56**(2), 39–44 (2018). https://doi.org/10.2525/ECB.56.39
6. Bocanegra-Herrera, C., Orejuela-Cabrera, J.: Cellular manufacturing system selection with multi-lean measures using optimization and simulation. Ingenieria y Universidad **21**(01), 7–25 (2017). https://doi.org/10.11144/JAVERIANA.IYU21-1.DCMS
7. Rewers, P., Diakun, J.: A Heijunka study for the production of standard parts included in a customized finished product. PLOS ONE **16**(12), e0260515 (2021). https://doi.org/10.1371/JOURNAL.PONE.0260515
8. Souza, J., et al.: Application of the single-minute exchange of die system to the CNC sector of a shoe mold company. Cogent Eng. **6** (2019). https://doi.org/10.1080/23311916.2019.1606376
9. Sahin, R., Kologlu, A.: A case study on reducing setup time using SMED on a turning line. J. Sci. **35**(1), 60–71 (2022). https://doi.org/10.35378/gujs.735969
10. Nedra, A., Néjib, S., Boubaker, J., Morched, C.: An integrated lean six sigma approach to modeling and simulation: a case study from clothing SME. Autex Res. J. **22**(3), 305–311 (2022). https://doi.org/10.2478/AUT-2021-0028
11. Nadja, D., Talib, D.: An approach to optimizing Kanban board workflow and shortening the project management plan. IEEE Trans. Eng. Manag. (99), 1–8 (2022). https://doi.org/10.1109/TEM.2021.3120984
12. Saini, M., Efimova, A., Chromjaková, F.: Value stream mapping of ocean imports containers: a process cycle efficiency perspective. Int. Sci. J. About Logist. **8**(4), 393–405 (2021). https://doi.org/10.22306/al.v8i4.245
13. Silva, A., Sá, J., Santos, G., Silva, F., Ferreira, L., Pereira, M.: A comparison of the application of the SMED methodology in two different cutting lines. Qual. Innov. Prosper. **25**(1) (2021). https://doi.org/10.12776/QIP.V25I1.1446
14. Magallanes, J., Camasi, C.: Costo de oportunidad del sector agrícola comercial exportador en el Perú, 1998–2017. Anales Científicos **82**(1), 33–41 (2021). https://doi.org/10.21704/ac.v82i1.1739

15. Juarez-Vite, A., Rivera, H., Corona, J., Montaño O.: Application of the SMED methodology through folding references for a bus manufacturing company. Int. J. Ind. Eng. Manag. 3(14), 232–243 (2023). https://doi.org/10.24867/IJIEM-2023-3-335
16. Barrientos-Ramos, N., Tapia-Cayetano, L., Maradiegue-Tuesta, F., Raymundo, C.: Lean manufacturing model of waste reduction using standardized work to reduce the defect rate in textile MSEs. In: Proceedings of the 18th LACCEI International Multi-Conference for Engineering, Education and Technology, Buenos Aires, Argentina (2020). https://doi.org/10.18687/LACCEI2020.1.1.356

Exploring the Appeal of Convenience Store Point Collection Systems Through the Evaluation Grid Method

Chu-Hsuan Lee[✉]

National United University, Miaoli, Taiwan
hsuan6389@gmail.com

Abstract. This research explores the factors contributing to the appeal of point collection activities in supermarkets and convenience stores. The study utilizes a systematic approach, including point accumulator definition, interview explanation, interviews, Evaluation Grid Method (EGM) compilation, and the creation of a hierarchical diagram illustrating point collection preferences. The EGM process begins by identifying original evaluation items and comparing product characteristics pairwise to understand their unique features. Subsequently, upper-level and lower-level concepts are identified, leading to the generation of individual models and tables for various tiers of point accumulators. In addition, interviews with five point accumulators focus on their preferences for point redemption products offered by major convenience store chains. These interviews aim to uncover the true appeal of point collection activities. To assess the importance of attributes in determining the attractiveness of point collection, we hold a questionnaire survey and analyze them using Hayashi's Quantitative Theory Type I. A total of 131 participants (65 males, 66 females, aged 19–40) were surveyed. Subsequently, we delved into the discussion of the five major factors related to point collection. The KJ method is employed to summarize the five key attractive factors of point redemption activities: "Sense of gain," "Earn the feeling," "Pleasure," "Healing feeling," and "Superiority." Notably, the sensation of receiving more value than expected emerges as a central theme, encompassing factors such as practicality and product versatility. This research enhances our understanding of the various elements that make point collection activities appealing to consumers. It sheds light on the significance of perceived value in these activities and how practicality and product versatility contribute to this perception. The findings offer valuable insights for marketing strategies in the retail industry.

Keywords: Evaluation Grid Method · KJ method · Hayashi's Quantitative Theory Type I · Miryoku Engineering

1 Introduction

1.1 Research Background and Motivation

With the products sold in different convenience stores exhibiting a high degree of homogeneity and negligible price differentials, a landscape of fierce market competition has emerged. In response to this challenging environment, major convenience store chains

S.-H. Sheu (Ed.): IEIM 2024, CCIS 2070, pp. 54–64, 2024.
https://doi.org/10.1007/978-3-031-56373-7_5

have strategically turned to incentivizing customer consumption through the implementation of a point-based system upon surpassing a predetermined spending threshold. This innovative marketing approach has swiftly evolved into a ubiquitous promotional strategy among chain convenience stores, reshaping consumer behavior.

Expanding beyond the traditional realm of rewarding customers with points, this strategy has taken on a multifaceted dimension. Convenience store chains now offer customers diverse options for utilizing their accumulated points, such as direct redemption for specific products or enjoying discounted purchases from affiliated partner companies. This evolution not only enhances the overall customer experience in the process of earning and redeeming points but also serves as a significant catalyst for an upswing in sales figures.

This study is dedicated to the meticulous collection and consolidation of the myriad point collection methods currently implemented by various convenience store operators. These methods can be broadly categorized into four distinct approaches: 1. Seasonal-themed product redemptions; 2. Point accumulation tied to the purchase of seasonal-themed products; 3. Irresistible buy-one-get-one-free promotions on select items; 4. Point accrual mechanisms for products available at different affiliated business outlets. By comprehensively examining these categories, the study aims to provide valuable insights into the evolving landscape of convenience store point collection strategies, shedding light on their impact on consumer engagement and market dynamics.

1.2 Purpose of Research

To comprehensively unravel the intricacies of consumers' engagement with convenience store point collection activities, this research employed a multi-faceted approach. In the initial phase, in-depth interviews were conducted utilizing the Evaluation Grid Method (EGM), a strategic tool designed to discern and categorize top, middle, and bottom-level factors influencing consumer behavior. The insights gleaned from this qualitative exploration laid the groundwork for the subsequent phase, wherein a meticulously crafted questionnaire was developed. The questionnaire, informed by the qualitative findings, served as the basis for a rigorous quantitative analysis, aimed at unraveling correlations and pinpointing specific factors that contribute to the magnetic appeal of convenience store point collection.

Furthermore, a nuanced examination was undertaken to compare the perceptions of consumers categorized into two distinct groups: those who actively partake in point collection activities at convenience stores and those who engage occasionally. This comparative analysis sought to unveil any discernible divergences in their perspectives on the allure of convenience store point collection. If disparities were identified, the subsequent step involved a meticulous investigation into the underlying factors contributing to these differing perceptions. Through this comprehensive methodology, the study aimed to provide a holistic understanding of the factors shaping consumers' attraction to convenience store point collection activities.

2 Literature Review

2.1 Kansei Engineering

With the advancement of technology and changes in lifestyle, consumers' considerations when purchasing products have expanded beyond functionality and practicality. The meanings of imagery and sensibility both refer to the psychological responses generated after receiving stimuli from objects. Kansei Engineering (KE) is an approach used to convert emotions and perceptions into specific product attributes [1]. Besides, it is a methodology that aligns customers' emotional responses with design elements to enhance the design in a way that best meets customers' preferences and expectations [2]. On the other hand, the term "Kansei" leans toward the philosophical realm, denoting a process of psychological functioning that starts with the relationship between people and objects. It defines the correlation between people's sensory perceptions, sensations, sensibilities, and motivations related to objects [3]. This emerging research field not only brings a new dimension to engineering but also, many research findings confirm its effectiveness in interpreting sensibilities.

2.2 Miryoku Engineering

Miryoku Engineering, initiated by Masato Ujigawa in 1991, advocates a preference-based design approach, which is a technical system for identifying attractiveness based on user preferences [4]. It also is a design approach grounded in understanding customer preferences, allowing product designers to engage directly and efficiently with the requirements of customers [5]. This approach also allows us to uncover the connection between preferences and attractiveness. In this study, the initial segment involves the use of the Evaluation Grid Method (EGM), while the subsequent part employs the Quantification Type I Method, as outlined below.

2.3 KJ Method

The KJ Method or KJ Technique, named after its creator Jiro Kawakita, is a technique used for generating ideas and setting priorities [6]. It's widely recognized and commonly employed in various contexts such as design, teamwork, retrospectives, and project meetings, making it one of the most popular variations of brainstorming. In brief, it is a systematic and integrative technique for discovering new meanings through creativity.

2.4 Evaluation Grid Method

The Evaluation Grid Method (EGM), introduced in 1986 by Japanese scholars Junichiro Sanui and Masao Inui as a part of Miryoku engineering, is rooted in psychologist Kelly's Repertory Grid Method [7]. It is a qualitative research technique that combines paired comparison and interpretive structural modeling to identify structural elements in-depth during interviews [8]. To grasp the evaluation criteria and the network structure of factors, EGM aims to extract the language used by consumers. It is a method to gain insight into how consumers assess the value of a product through in-depth individual investigations.

In short, it primarily conducts interviews within deeply engaged ethnic groups, comparing various characteristics derived from actual user behavior cases to extract authentic evaluations and opinions [9]. Within the realm of attractiveness engineering research, EGM offers a tangible and systematic approach to analyze the factors contributing to the attractiveness of objects. To comprehend the user's perception of the appeal of objects, we employ in-depth interviews. Stimuli are presented within specific categories under a given theme, allowing for a comparison of the subjects' preferences and resulting in distinct and evident impressions for the subjects. During this investigative process, the analysis delves into what consumers perceive and where they place value, all accomplished through interviews. Particularly, when it comes to the hierarchical structure of semantic visualization, it allows us to visualize and concretize consumers' value structures, progressing from the abstract to the specific, by organizing evaluation elements into top, middle, and bottom-level factors.

2.5 Hayashi's Quantitative Theory Type I

This research adopts I-type quantification as a robust analytical tool for evaluating the allure of point collection. The utilization of this method offers a comprehensive assessment of attraction significance, allowing for the measurement and quantification of the importance attributed to raw ratings, letter-up, and letter-down items. I-type quantification, particularly when applied in conjunction with multiple linear regression techniques as outlined in Hayashi's Quantitative Theory Type I [10], serves as a powerful statistical tool to predict the relationships between response values and categorical values.

In the realm of consumer experience, the versatility of Hayashi's Quantitative Theory Type I becomes apparent as it facilitates the determination of weights assigned to various factors based on user preferences [11, 12]. This not only enhances the precision of the analysis but also provides a nuanced understanding of how different factors contribute to the overall appeal of point collection activities. By employing this methodological approach, the study aims to unravel the intricate dynamics that underlie consumer perceptions and preferences in the context of point collection, offering valuable insights into the factors that significantly influence the attractiveness of such programs. The application of I-type quantification ensures a rigorous and data-driven exploration, contributing to the robustness of the study's findings.

3 Research Methods

3.1 Research Process

The research steps, in sequential order, include point accumulator definition, interview explanation, interviews, EGM compilation, and the hierarchical diagram of point collection preferences by EGM.

There are three criteria for defining a 'Point Accumulator': 1. Those who have accumulated points and redeemed rewards five times or more, 2. Those who have engaged in point accumulation activities within the past six months, 3. Those who can name the recent redemption products offered by convenience stores. Based on these criteria,

EGM data is compiled. Initially, we identify the original evaluation items, comparing the characteristics of products pairwise to understand their winning attributes. Subsequently, we identify upper-level and lower-level concepts, and finally, we generate individual models and compile tables for the top, middle, and bottom tiers. Next, in terms of point redemption products (specifically, those from 7-Eleven, Family Mart, and Hi Life), after interviewing five point accumulators, we aim to understand their preferences for these products. Through these in-depth interviews, we seek to uncover the true allure of point collection activities.

The next step is to use the KJ method to summarize the five attractive factors of the point redemption activity, which are, "Sense of gain", "Earn the feeling", "Pleasure", "Healing felling", and "Superiority". Among these, the sensation of getting more value than expected includes the highest number of middle-tier factors, which are: practicality, and product versatility. From this, it can be deduced that point accumulators largely consider the perception of getting more value than expected as the primary allure of collection activities, with practicality and product versatility being the key contributing factors to this sense of added value.

3.2   The Hierarchical Diagram of Collecting Points

Upon completion of the aforementioned research steps, a comprehensive evaluation structure for all point accumulators is established. Figure 1, the hierarchical diagram of point collection preferences by EGM, serves as a visual representation of the intricate relationships and hierarchies within the accumulated data. This diagram encapsulates the essence of point accumulation preferences, providing a systematic overview of the factors influencing participants' choices and priorities in the context of reward redemption.

To delve deeper into the utilization of Fig. 1, it acts as a powerful tool for synthesizing the diverse perspectives obtained through preference interviews conducted by EGM. Each node in the hierarchical diagram corresponds to a specific aspect of point collection, reflecting the nuances uncovered during the interviews with point accumulators. The connections and clustering within the diagram highlight the interdependencies and correlations among different preferences, shedding light on the underlying patterns that contribute to the allure of collecting points.

The data analysis was employed using the KJ method to organize findings into five major categories: 1. Sense of gain, 2. Earn the feeling, 3. Pleasure, 4. Healing feeling, 5. Sense of surprise. Each of these categories is explained below:

- Sense of gain
  In this category, the top-level factor is the sense of getting value for money. The middle-level factors, in sequential order, include Practicality (Coffee), Practicality (Product), and product functionality diversity. The bottom-level factors consist of shopping twice a week, styluses and decorations.
- Earn the feeling
  In this category, the top-level factor is Earn the feeling. The middle-level factors include the flexibility to choose the redemption method and a low cumulative price threshold. The bottom-level factors involve the option to make additional purchases with a supplemental fee or entirely redeem with points, as well as schemes offering

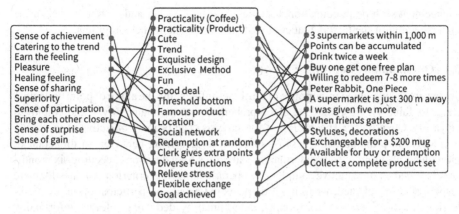

Fig. 1. The hierarchical diagram of point collection preferences by EGM

buy one, get one free, allowing for savings of several tens of dollars, and point can be accumulated.

- Pleasure

  At the top-level factor of this category is the sensation of pleasure. The middle-level factor includes cute cartoon designs and products that are entertaining and fun. The bottom-level factor consists of styluses and decorations.

- Healing feeling

  In this category, the top-level is the sense of healing. The middle-level factors include cute cartoon designs and achieving a goal by completing a set. The bottom-level factors consist of styluses and decorations, as well as completing a full set of products.

- Sense of surprise

  In the realm of a sense of superiority, the middle-level factors include items that cannot be purchased unless exchanged and products associated with renowned brands. The bottom-level factors involve a willingness to exchange seven to eight times if the desired product is not obtained through redemption.

3.3   Preparation for Questionnaire

Following this, a meticulous analysis employing the KJ method was undertaken, resulting in the identification of five pivotal factors extracted from the initial evaluation criteria. These factors are succinctly labeled as "Sense of gain," "Earn the feeling," "Pleasure," "Healing feeling," and "Superiority". By precisely delineating attribute items and categories grounded in a nuanced comprehension of these factors, the subsequent iteration of questionnaires is anticipated to be more resilient and efficient in encapsulating the intricacies of respondents' experiences.

The deliberate categorization of these factors serves as a strategic refinement, providing a more granular lens through which to interpret participant responses. This methodological enhancement not only facilitates a more thorough exploration of consumer perceptions but also augments the reliability and validity of the data collected. As a result, the study is better positioned to unravel the multifaceted dimensions of individuals' engagement with point collection activities, offering a richer and more comprehensive understanding of the factors that contribute to the appeal of such programs.

3.4 Questionnaire Survey

To discern the weighted importance of attributes influencing the allure of point collection (refer to Table 1), we conducted a comprehensive questionnaire survey and employed Hayashi's Quantitative Theory Type I for analysis. The primary objective of this study is to delve into the attractiveness of supermarket gathering behavior. In this context, individuals abstaining from supermarket gatherings for a duration exceeding six months are classified as unattractive samples. To operationalize this criterion, 46 questionnaire responses were selected, wherein participants indicated non-participation in point collection activities for over six months, determined by deducting relevant information from the fifth questionnaire item. This meticulous process resulted in a final dataset comprising 85 questionnaires for analysis.

The subsequent phase of this study involves a detailed examination and comparison of results. The aim is to ascertain whether the magnitude of individuals drawn to supermarket shopping activities has a discernible impact on the overall attractiveness of these endeavors. Following this quantitative analysis, the study pivots towards an in-depth exploration of the five major factors intricately linked to point collection. By dissecting and understanding these factors, the research seeks to provide nuanced insights into the dynamics shaping the appeal of supermarket gathering behavior, thereby contributing to a more comprehensive understanding of consumer preferences in the retail landscape.

Table 1. The best 5 of letter-up from hierarchical diagram.

Factor	Reason
Sense of gain	Practicality (Coffee)
	Practicality (Product)
	Diverse Functions
Earn the feeling	Flexible exchange
	Threshold bottom
Pleasure	Cute cartoon shape
	Product interesting and fun
Healing felling	Cute cartoon shape
	Collect a set to achieve the goals
Superiority	Gift's Exclusive Redemption Method
	Famous product

4   Results

4.1   Results of Hayashi's Quantitative Theory Type I

This study aims to meticulously conduct a separate analysis of each discernible factor and subsequently scrutinize and compare the obtained results. The specific factors under comprehensive investigation encompass "Sense of gain" (Practicality, Diverse Functions), "Earn the feeling" (Flexible exchange, Threshold bottom), "Pleasure" (Cute cartoon shape, Product interesting and fun), "Healing feeling" (Cute cartoon shape, Collect a set to achieve goals), and "Superiority" (Gift's Exclusive Redemption Method, Famous product). Each of these designated factors is meticulously treated as individual items, emphasizing a granular examination, while the lower level effectively represents the overarching category.

The ensuing discussion delves into the analysis results based on the Partial Correlation Coefficient (as depicted in Table 2). Among the 131 samples, the highest item score, reaching 0.0894, pertains to the factors of pleasure (Cute cartoon shape, Product interesting and fun), excluding those who haven't engaged in collection activities for more than half a year. Remarkably, the 85 samples post-exclusion similarly registered the highest scores in the pleasure components. This suggests a consensus among participants, whether avid or occasional collectors, that the elements contributing to pleasure are the primary factors influencing the allure of point collection in supermarkets.

While the lowest recorded score pertained to the components of the healing feeling (Cute cartoon shape, Collect a set to achieve goals) with 131 samples (0.0083), a conspicuous and significant distinction is evident when comparing the results of the 85 samples to the more extensive sample pool of 131. Consequently, this research unequivocally asserts that individuals actively participating in collection activities overwhelmingly concur that the healing component items wield a certain level of influence on the overall charm of supermarket points. This nuanced and in-depth exploration affords a more profound understanding of the intricate and multifaceted factors at play in shaping consumer perceptions and preferences within the realm of supermarket point collection.

Table 2 highlights that the pivotal element attracting consumers to engage in point collection activities is the sensation of "Pleasure". Additionally, the category exerting a significant impact on the "Sense of gain" is identified as the practicality of redemption items. Examining the results, it is evident that 131 respondents (0.7797826) and 85 respondents (0.170588) reveal a notable gap between the two perspectives. Consequently, it is inferred that individuals less inclined to frequent exchange activities concur on the practicality of exchange items being the primary factor influencing supermarket point collection activities.

In terms of the factor labeled "Earn the feeling," both frequent and occasional redeemers, as indicated by the sample results of 131 (0.303226) and 85 (0.03225) people, align in their agreement that the buy one, get one free program significantly contributes to the sense of earning. While "Pleasure" emerges as the paramount factor during supermarket point redemption, the category with the most considerable impact, namely the brand of the product, only registers at 0.145762. There exists an intriguing phenomenon to explore within the "Healing feeling" factor. In the category of collecting a complete

Table 2. The category scores 5 for items of factors.

Items	Categories	CS		PCC	
		131	85	131	85
Factor 1: Sense of gain					
Practicality	Practicality of redemption items	0.7797826	0.170588	0.0568	0.0567
Diverse Functions	Product functions are diverse	−0.431765	−0.113725		
Factor 2: Earn the feeling					
-Flexible exchange	Points or purchase price choice	0.217500	0.210200	0.0836	0.0835
	Buy one get one free plan	0.303226	0.03225		
Threshold bottom	Easily earn points	−0.19333	0,009210		
Factor 3: Pleasure					
Cute cartoon shape	Nature of the product	−0.33076	−0.320769	*0.0894	0.0894
Product interesting and fun	Brand of the product	0.145762	0.143762		
Factor 4: Healing felling					
Cute cartoon shape	Brand of the product	0.098148	0.098148	0.0083	0.0727
	Nature of the product	−0.17096	−0.170967		
Collect a set to achieve the goals	Collect a complete set of products	0.41176	−0.10294		
Factor 5: Superiority					
Gift's Exclusive Redemption Method	Well-known brand	0.112903	0.112903	0.0775	0.0755
Famous product	Exchange Again for Preferred Product	−0.30434	−0.304347		

set of products, the analysis reveals a score of 0.41176 for the 131-sample group, contrasting with a −0.10294 for the 85-sample group. This substantial difference suggests that respondents who infrequently redeem points may significantly influence the analysis outcomes regarding the factors that contribute to the appeal of point collection.

Both sets of analyses underscore the significance of the "Superiority" category, particularly concerning well-known brands, as the primary influencing factor shaping respondents' willingness to collect points. This insight enriches our understanding of consumer behavior and preferences in the context of supermarket point collection activities.

4.2 Discussion

In analyzing the numerical results obtained through Hayashi's Quantitative Theory Type I calculations, it was observed that the values for 131 or 85 questionnaires were slightly similar in terms of partial correlation coefficients or category scores. It is inferred that the possible reason for this similarity is that regardless of whether individuals are enthusiastic participants in the redemption activity, they share a consistent perspective on the constituent elements proposed in this study. For example, it appears that most people generally favor the idea of getting value for their money or feeling like they've gained something, which leads to a common consensus even among those who don't frequently engage in the activity. Additionally, it should be noted that nearly 55% of the respondents in this study were students, which may have contributed to the similarity in response patterns.

As for the potential reasons for the observed differences in values, this study bifurcated the questionnaires into two segments. The initial part encompassed a total of 131 questionnaires, whereas the latter part consisted of a total of 85 questionnaires after excluding infrequent point collectors. Both segments underwent analysis and comparison, elucidating disparities in specific items. Notably, in the constituent items associated with the "Healing feeling" (such as a cute cartoon shape and collecting a set to achieve goals), there were substantial differences in the results between distinct sample groups. Consequently, this research posits that individuals who actively engage in redemption activities generally concur that the "Healing feeling" constituent items wield a certain level of influence on the allure of convenience store point collection. From the aforementioned arguments, it can be inferred that frequent participants in this activity harbor divergent preferences and viewpoints on certain items compared to their less frequent counterparts.

5 Conclusion

In conclusion, this study aimed to investigate the factors that contribute to the attractiveness of supermarket point collection activities among consumers. We conducted a comprehensive analysis of various factors related to point collection, including "Sense of gain," "Earn the feeling," "Pleasure," "Healing feeling," and "Superiority." Each of these factors was further divided into constituent elements, providing a detailed understanding of the components that drive the appeal of point collection. Our analysis, based on Hayashi's Quantitative Theory Type I, revealed interesting insights. The most significant factor that attracts consumers to participate in point collection activities is the feeling of "Pleasure". Regardless of whether participants are enthusiastic or occasional collectors, they unanimously agree that the elements contributing to pleasure are the primary factors that enhance the charm of point collection at supermarkets.

Additionally, we observed differences in responses between frequent and infrequent participants in certain factors. For instance, the "Healing feeling" category showed significant variations, indicating that individuals actively engaged in collection activities tend to attribute greater importance to this aspect. Furthermore, we identified that the "Superiority" category, particularly the well-known brand of the product, significantly influences consumers' willingness to collect points.

In summary, this study sheds light on the diverse factors that drive consumers' engagement in supermarket point collection activities. It highlights the importance of the "Pleasure" factor while recognizing that individual preferences may vary based on their level of participation. These findings contribute to a deeper understanding of the appeal of point collection in supermarkets and offer valuable insights for marketing strategies in the retail industry.

References

1. Nagamachi, M., Ishihara, S.: Kansei engineering. In: Proceedings of the 13th International Conference on Applied Human Factors and Ergonomics (AHFE 2022), New York, USA, 24–28 July 2022 (2022)
2. Lin, S., Shen, T., Guo, W.: Evolution and emerging trends of Kansei engineering: a visual analysis based on CiteSpace. IEEE Access **9**, 11181–111202 (2021). https://doi.org/10.1109/ACCESS.2021.3102606
3. Kelly, G.A.: The Psychology of Personal Constructs. Norton, New York (1955)
4. Sanui, J.: Visualization of users' requirements: introduction of the evaluation grid method. In: Proceedings of the 3rd Design & Decision Support Systems in Architecture & Urban Planning Conference, vol. 1, pp. 365–374 (1996)
5. Hsieh, M.Y., Wang, I.C.: Combining Miryoku engineering and evaluation grid method explore wireless headphone design appealing factors. In: 2022 25th International Conference on Mechatronics Technology (ICMT), Kaohsiung, Taiwan, pp. 1–4 (2022). https://doi.org/10.1109/ICMT56556.2022.9997718
6. Miryoku Engineering Forum: Miryoku Engineering. Kaibundo (Japanese), Tokyo (1992)
7. Kawakita, J.: The Original KJ Method, Revised edn. Kawakita Research Institute (1991)
8. Nagamachi, M.: Kansei engineering: a new ergonomic consumer-oriented technology for product development. Int. J. Ind. Ergon. **15**(1), 3–11 (1995)
9. Shen, Q., Chen, C.C., Wu, S.M.: Research of the attractiveness factors of MOBA mobile games based on evaluation grid method. J. Phys. Conf. Ser. **1325** (2019). https://doi.org/10.1088/1742-6596/1325/1/012004
10. Hayashi, C.: On the quantification of qualitative data from the mathematico statistical point of view. Ann. Inst. Stat. Math. **2** (1950)
11. Iwabuchi C., et al.: Data Management and Analysis by Yourself, pp. pp180–185. Humura Publishing, Japan (2001)
12. Sugiyama, K., et al.: The Basic for Survey and Analysis by Excel, pp. 51–62. Kaibundo Publishing, Japan (1996)

Stockout Reduction Using Forecasting Methods, the EOQ Model and a Safety Stock in a Peruvian SME in the Commercial Sector

Ariana Alisson Borja-Gonzales[ID], Alexandra Beatriz Perez-Soto[ID],
and Alberto Flores-Perez[(✉)] [ID]

Facultad de Ingeniería, Universidad de Lima, Lima, Peru
{20171907,20201623}@aloe.ulima.edu.pe, Alflores@ulima.edu.pe

Abstract. Effective inventory management is a constant challenge for SMEs. Stockouts represent a significant obstacle that affects the operability and profitability of these companies. The aim of this study is to develop an affordable and comprehensible solution to reduce stockout in a Peruvian company in the commercial sector. This article presents an improvement proposal leveraging forecasting methods (Holt-Winters and SARIMA), the EOQ model and a safety stock. To evaluate the efficacy of our model, simulations were conducted using Microsoft Excel, followed by the assessment of three key indicators: stockout, bias and inventory turnover. The results indicate a substantial decrease in stockouts by 75%. An economic evaluation demonstrated the prospective profitability of the proposal, as it yielded a Net Present Value (NPV) of 9 031 USD, an Internal Rate of Return (IRR) of 92% and a payback period of 59 days. Subsequent sensitivity analysis confirmed profitability even under adverse scenarios. These findings, coupled with the model's adaptability and the generalizability of conditions, imply broader prospects for the implementation of this proposal, extending its relevance to diverse industries and countries.

Keywords: Forecast · Holt-Winters · SARIMA · EOQ · Safety Stock · Stockout

1 Introduction

Since 1990, more than a billion people have benefited from business practices and today the market share has been quantitatively important in the market, maintaining a market share of more than 90% over the last few years [1], which indicates that it is crucial for today's economic progress. Within this area, even though there are tools to solve it, stockouts persist, i.e., when a company runs out of inventory, in simple words, it means that an item is out of stock and a customer's order cannot be fulfilled [2]. Such a situation mainly arises in small and medium-sized enterprises (SMEs) due to inaccurate demand estimation, as they tend to rely on sales experience from previous years considering that the same sales of previous months will be repeated in the next months of the following years, without using mathematical calculations, while consolidated companies resort to expensive advanced software. While neither of these methods guarantees an accurate

© The Author(s), under exclusive license to Springer Nature Switzerland AG 2024
S.-H. Sheu (Ed.): IEIM 2024, CCIS 2070, pp. 65–75, 2024.
https://doi.org/10.1007/978-3-031-56373-7_6

prediction of demand, their application often yields favorable results with minimal margin of running out of inventory. These companies, lacking scientific forecasting methods, do not have the resources or expertise to make an accurate estimation of demand. In the following sections, we will delve into the specific challenges faced by these small and medium-sized enterprises regarding stockouts.

This situation is worrying since SMEs constitute 99% of all companies and generate 77% of employment [3]. Therefore, this work is justified by the importance of avoiding stockouts, not only to prevent immediate economic losses, but also to avoid negative impacts on customer relations, company prestige and both social and economic consequences. Therefore, a comprehensive analysis of several case studies was conducted to validate the effectiveness of tools for forecasting demand and improving inventory management. One of these case studies mentions the development of a forecasting method to determine electricity demand in Spain during holidays one day in advance [4]. This method is based on multiple seasonal Holt-Winters models and was compared with four benchmark methods (ANN, Bagged Regression Trees (BRT), Exponential Smoothing with Box Cox Transform and ARMA residual modelling (BATS), and the Trigonometric Seasonal version of the latter (TBATS)), as well as with a specific method for anomalous demands: the Rule-Based SARMA (RB-SARMA). The results indicated that the new approach outperforms the usual methods in accuracy, reducing the forecasting error from 9.5% to less than 5% during holidays. In another paper, they implemented EOQ for inventory management improvement at a drug manufacturing plant in Montreal, generating annual savings of 110 800 USD [5]. The algorithm helped management decide to invest in larger production equipment by handling peaks in demand for Class A items. After implementation, inventories were significantly reduced without compromising high service levels. Hence, the feasibility of implementing these tools in medium-sized companies in the commercial sector is confirmed. Therefore, considering the characteristics of small and medium-sized companies, it is proposed to apply the Holt-Winters forecasting method and the SARIMA model on a specific product, which is the one that generates the most sales for the company, as well as to implement the EOQ and a safety stock, due to their low implementation cost and ease of application, also considering the characteristics of such a company such as its seasonality and inventory costs.

Finally, with the aim of offering a proposal for improvement, the imperative need arises to carry out this research, which is structured in various sections, including introduction, state of art, contribution, validation, discussion, and conclusions.

2 State of Art

2.1 Forecasting Methods to Predict Demand

To predict the demand, two forecasting techniques were employed. The first one is the Holt-Winters method, also called as triple exponential smoothing model, used for short-term forecasting with seasonal and trend patterns [6]. The effectiveness of this tool has been proved in several sectors, such as energy, where its application in a case study in Spain reduced the forecasting error from 9.5% to under 5% [7]. The second one is the SARIMA model, which incorporates seasonal components to handle time series data with recurring patterns at fixed intervals, such as daily, weekly, or yearly cycles [8]. This model is remarkably versatile and accurate as well, as can be seen in a case study in the hotel sector in Thailand, where out of 4 different forecasting models, SARIMA provided the best accuracy [9].

2.2 The EOQ Model to Optimize Inventory Management

To improve inventory management, the economic order quantity (EOQ), also known as optimal lot or economic order lot, was used. This tool has been shown in other companies to minimize the total cost of inventories and to avoid stockouts, as it optimizes replenishment, avoiding frequent orders resulting in shortages and thus more efficient inventory management. In a case study, for example, analysis of the results concluded that, had the EOQ method been employed, there would have been a saving of Rp 10 800 000 with a percentage of 54.18% in the cost of ordering and storing raw materials [10]. Furthermore, in another study conducted in an Italian company in the railway industry, the implementation of Economic Order Quantity (EOQ) and Economic Order Interval (EOI) was simulated and compared across 53 items. The results demonstrated superior outcomes in terms of economic savings and inventory quantity compared to the company's current management policy. Specifically, the EOQ policy was selected for implementation throughout the warehouse, emphasizing its effectiveness and practical applicability in diverse business environments. This adoption not only led to significant cost savings and improved inventory levels but also contributed to enhanced overall operational efficiency [11]. This model is effective in optimizing inventory management costs, adapting to diverse business environments. Its simplicity of implementation and ability to handle variations in demand make it accessible and efficient, contributing to improve the operational efficiency of companies.

2.3 A Safety Stock to Prevent Supply Interruptions

A safety stock is an extra inventory held in the warehouse to deal with unforeseen events related to changes in demand or supplier delays [12]. It has been used in several compa- nies and has made it possible to avoid stock-outs in a short period of time, as was the case in Brazilian hospitals during the COVID 19 pandemic, where this tool made it possible to adopt strategies to control and maintain stocks at safe levels according to demand, thus mitigating the risk of stock-outs [13]. In another case study, the company imple- mented predictive safety stock for each journey, carefully considering uncertainties in demand and sailing time. This strategic approach proved to be highly effective, resulting in a significant reduction in the risk of shortages. The company achieved an 81.44% improvement in safety stock efficiency through a heterogeneous approach compared to homogeneous conditions for different route distances. This specific implementation has allowed the company to optimize its fuel supply and achieve notable cost savings, emphasizing the practical applicability of safety stock in diverse business environments [14]. Thanks to this additional buffer, an adequate level of stock can be maintained dur- ing these situations, providing a safety margin to meet demand without disruption and facilitating smoother and more efficient operations.

3 Contribution

3.1 A Model Basis

After identifying the root causes of the stockout problem, evaluating the tools available and conducting an exhaustive bibliographical review, we concluded that the best improvement tools considering our context were forecasting methods (Holt-Winters and SARIMA) for improvement of the current demand determination methods, EOQ for optimization of the inventory management, and a safety stock to directly reduce the risk of stockout. Considering this, a comparison matrix with the articles that contributed the most scientifically and the tools suggested is presented in Table 1.

Table 1. Comparison matrix of the proposal components vs the state of the art.

Articles	Component			
	Improve demand determination methods	Improve demand determination methods	Optimize inventory management	Reduce the risk of stockout
Trull, O., García-Díaz, J.C., & Troncoso, A. (2021)	Holt-Winters			
Hwandee, O., & Phumchusri, N. (2020)		SARIMA		
Christifan, A. & Gonzali, L. (2020)			EOQ	
Gonçalves, J. (2020)				Safety stock
Proposal	Forecasting method (Holt-Winters)	Forecasting method (SARIMA)	EOQ	Safety stock

3.2 Proposed Model

The proposed model consists of 3 tools: Forecasting methods, the EOQ model and a safety stock. These will work sequentially, as the outcome from one process informs the subsequent one.

3.3 Model Components

Component 1: Estimation
As the first component of our model, demand estimations through proper engineering methods are implemented as a replacement for the previous non-mathematical methods based just in previous experience. We decided to use two different forecasting methods so we could simulate, based on the historical data, which would have calculated better the demand, and work with said method permanently.

Component 2: Planning
The second phase involves the application of the EOQ tool, which reduces costs, as orders would be placed monthly and no longer weekly, also considering the forecast demand that has been found in component 1. This calculation made by the companies represents the ideal size of their order, allowing them to meet demand without overspending [15]. This reduces the risks associated with inadequate stock levels and contributes to more efficient inventory management. For this purpose, the following formula is applied:

$$EOQ = \sqrt{\frac{2*D*S}{H}}$$

D: annual demand
S: order cost per lot
H: unit cost of storage

Component 3: Prevention
The third phase consists of implementing a safety stock, which employs the use of a distribution table to find the value of Z. This reduces the risk of running out of stock and, while improving operational efficiency, ensures an adequate response to changes in demand by optimizing stock levels.

$$SS = Z * \sigma\, total$$

$$\sigma\ total: \sqrt{(\sigma_1^2 + (\sigma_2^2)}$$

Z: Constant, dependent on the expected Level of Service
σ_1^2: Cumulative deviation of demand during lead time (Lead time)
σ_2^2: Lead time deviation expressed in units from demand demand (Fig. 1).

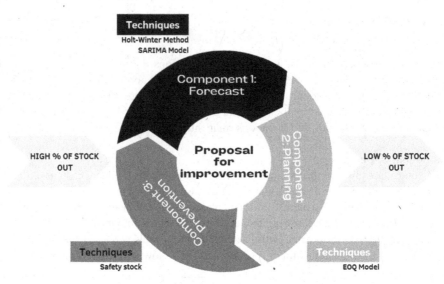

Fig. 1. Proposed method to reduce the stockout.

3.4 Indicators

To evaluate the effectiveness and measure the improvements resulting from the implementation of the proposed models, the following indicators were selected.

- **Stockout rate:** Measures the frequency with which the company runs out of inventory before replenishment, assessing the ability to meet demand, expressed as a percentage. Objective: Reduce the incidence of times when the company experiences stockouts, ensuring a continuous response to customer demand and preventing lost sales.

$$Stockouts = \frac{number\ of\ unsatisfied\ orders}{total\ number\ of\ orders} * 100$$

- **BIAS rate:** Assesses the direction and consistency of deviations between a forecast and actual values, providing information on the accuracy of the model by indicating whether it tends to overestimate or underestimate the forecasted variable. When the BIAS is close to zero, the forecast is considered unbiased, while positive or negative values suggest a tendency to overestimate or underestimate, respectively.
Objective: Assess the accuracy and trend of a forecasting model, indicating whether it tends to overestimate or underestimate the variable. It seeks to improve the quality of forecasts by identifying systematic patterns of error.

$$BIAS = \frac{1}{num\_observations} * \sum_{i=1}^{num\_observation} FVPi - AVPi$$

- **Inventory turnover ratio:** It is calculated to assess how efficiently a company sells and replenishes its inventory in relation to total sales.

Objective: Optimize resources, minimize obsolescence, and improve profitability by meeting customer demand efficiently.

$$Turnover = \frac{sales\ for\ the\ period}{average\ inventory\ for\ the\ period} * 100$$

3.5 Validation

Initial Diagnosis
After a thorough analysis of the company's historical data, the average stockout calculated was 60.61%, which costs the company 113 100 USD in annual revenue. Since the industry standard was 8%, there was a technical gap of 52.61%, which indicated a serious problem. The implementation of several diagnostic techniques revealed that the reasons for the stockout were the following: estimates below demand (40.98%), not having a safety stock (31.94%) and supplier delays (27.08%).

Validation Design and Comparison with the Initial Diagnosis
The software selected for our validation was Microsoft Excel, for its user-friendly interface. Based on historical sales data from previous periods, the formulas of the tools indicated in item 3.3 were used to forecast the demand, to estimate the EOQ and subsequently the calculation of safety stock. All this was developed to evaluate both the variation with respect to previous periods and for subsequent months. To evaluate this variation with respect to previous years, a comparison was made using indicators such as the Stockout rate, BIAS and Inventory turnover, which provided us with key information on the accuracy of forecasts, inventory management efficiency and the system's capacity to avoid stockouts. If the percentage of those indicators was closer to 0, it indicated that there was a close alignment between predictions and reality, i.e., they were in sync.

Improvement-Proposal simulation
To validate the accuracy and effectiveness of the proposed model, a hypothetical scenario was simulated: for the best-selling product at the company, we calculated how, under the same circumstances of the period 2022, would the indicators have changed, had we used the data obtain by the improvement tools.

After developing each component through Microsoft Excel with the formulas mentioned in Sect. 3.3, we concluded that implementing the proposal would have reduced the stockout problem in the company by 75.13%.

As per the indicators evaluated, the following table shows a comparison between before and after the model was implemented (Table 2).

Furthermore, besides analyzing the performance of the proposed model and its indicators, the economic feasibility of the project's implementation was evaluated as well. To do so, we considered 3 variables. The first one was investment: the budget to train the workers, the time needed to do so, and the buying of the forecasting software. The second one was the income, which we considered the missed sales opportunity. The third and final variable was the expense, for which was considered the cost of the product itself and all its logistical costs (storage, transportation, and labor). With this information, we calculated an Income Statement for the 2022 period. An above-average inventory

Table 2. Indicators before and after implementing the proposed model.

Indicator	Reality	Improved
Stockout rate	55.43%	13.79%
BIAS rate	100%	42.21%
Inventory turnover ratio	47,43%	87,80%

turnover signals sales efficiency and dynamic inventory management, indicating that the company converts its inventory into revenue efficiently, decreasing obsolescence risks and improving cash generation (Table 3).

Table 3. Simulated Income Statement for the 2022 period, per month and in USD,

	0	1	2	3	4	5	6	7	8	9	10	11	12
Revenue		4 680	5 460	7 800	5 460	3 705	9 360	7 800	15 600	11 700	8 580	14 040	6 240
Cost of goods sold	−1 249	−3 009	−3 509	−5 037	−3 509	−2 377	−6 037	−5 009	−11 009	−7 537	−5 509	−9 009	−4 037
Gross profit	−1 249	1 671	1 951	2 763	1 951	1 328	3 323	2 791	5 591	4 163	3 071	5 031	2203
Taxes	−369	493	576	815	576	392	980	823	1 649	1 228	906	1 484	650
Net income	−1 617	1 178	1 375	1 948	1 375	936	2343	1 968	3 942	2 935	2 165	3 547	1 553

Since the Net Present Value (NPV) was positive (9 679 USD), the Internal Rate of Return (92%) exceeded the Cost of Capital (16.5%), the benefit-cost ratio (B/C) was 10.43 and the payback period were 59 days, we could affirm that the project was economically profitable.

In addition to that, a sensitivity analysis was developed via Risk Simulator using 10 000 interactions to study the best and worst possible outcomes from the implementation of our proposed model. Regarding the NPV, it revealed that even in the least favorable conditions, the project is still profitable. As for the IRR, its simulation showed that the variable that affects the most is the investment, and the least is fixed costs (Table 4).

Table 4. Summary of NPV and IRR for each scenario.

Indicators	Scenarios		
	Optimistic	Baseline	Pessimistic
NPV	9 766 USD	8 900 USD	7 911 USD
IRR	139,97%	91,92%	65,36%

4   Discussion

The obtained results unveil a pronounced superiority of the Holt-Winters model over SARIMA in forecast accuracy for inventory management, as evidenced by a significant 70% reduction in stockouts. The implementation of EOQ proved effective by placing monthly orders in single batches, optimizing timelines, and preventing shortages. Additionally, the introduction of a 5% safety stock has been essential in mitigating unforeseen fluctuations in demand.

While comparing our results with previous research, the importance of selecting the most suitable forecasting method for each context becomes evident. Although in certain scenarios the SARIMA method may outperform Holt-Winters, as demonstrated in studies related to hotel occupancy [16], for our inventory management, the Holt-Winters model proved to be the most accurate option. The adaptability of Holt-Winters to our real data, with only a 5% variation compared to SARIMA, underscores its suitability. Furthermore, prior studies indicate that the application of Holt-Winters in the electric consumption sector has resulted in a significant reduction in forecast error, dropping from 9.5% to less than 5% [17]. These findings emphasize the importance of assessing and selecting the forecast model that best fits the specific characteristics of each application, confirming the effectiveness of Holt-Winters in our inventory management.

On one hand, despite a significant reduction in stockouts, a slight discrepancy is attributed to the nuances in the company's sales policy tracking, potentially leading to overstock situations. This unexpected outcome underscores the importance of considering both internal dynamics and external variables in inventory management, emphasizing the complexity of the business environment and the need for adaptability in forecasting strategies.

Furthermore, the reduction in stockouts not only benefits businesses but also yields positive social impacts by ensuring the constant availability of essential tools, preventing shortages, meeting the needs of various sectors, enhancing individual customer satisfaction, and reducing wait times. This, in turn, provides a positive experience for the community.

On the other hand, the implementation of the proposed tools could face limitations due to the quality of historical data, impacting both EOQ and safety stock with an inaccurate demand forecast. Variability in a dynamic business environment, adaptation to changing policies, and unpredictable external factors such as delivery delays from suppliers also pose challenges. These limitations underscore the need for constant vigilance and adaptability to ensure effectiveness in inventory management for the SME.

Finally, it is recommended to establish a continuous monitoring system, provide staff training, and maintain flexibility in commercial policies. Setting clear goals, effective collaboration with suppliers, and considering additional technologies are essential. Conducting periodic reviews and collecting customer feedback will contribute to continuous improvement, thereby strengthening inventory management in the SME within the commercial sector.

5 Conclusions

In order to determine the feasibility of our proposal, two main subjects were analyzed: the accuracy of the improvement tools, and the profitability of its application. As for the improvement tools, for the first component we determined that the Holt-Winters method provided more precise results over SARIMA, therefore the data would be forecasted with the first method. From the second and third component can be learned that, while it ultimately accomplished the goal of preventing stockout and reducing costs, there were some values higher than expected, which were attributed to the variability of the company's sales' policies,

Overall, after analyzing the discrepancies in certain values' accuracy from the 3 tools proposed, it was determined that the tool itself was not causing this issue, but rather the changing policies from the company.

As for the profitability of its application, the improvement proposal generated significant revenue considering the inexpensive investment required (less than 2 000 USD). The sensitivity analysis conducted also showed that, even in the most pessimistic scenario, the model proved to still be profitable.

Based on the effectiveness of the tools suggested and its financial success, we conclude that the improvement proposal could be successful in reducing the stockout in the company.

References

1. Banco Mundial: Panorama General del comercio (2021)
2. SimpliRoute. https://simpliroute.com/es/blog/5-formas-para-prevenir-los-quiebres-de-stock
3. Global Business and Economics Research Centre: International comparison of the contribution of MSMEs to the economy, pp 1–2 (2019)
4. Dieudonné, N.T., Armel, T.K.F., Vidal, A.K.C., René, T.: Forecasting irregular seasonal power consumption. An application to a hot-dip galvanizing process. Appl. Sci., 75 (2020)
5. Ouellet, R., Roy, J., Cardinal, C., Rosconi, Y.: EOQ application in a pharmaceutical environment: a case study (1982)
6. Swapnarekha, H., Behera, H.S., Nayak, J., Naik, B., Kumar, P.S.: Multiplicative holts winter model for trend analysis and forecasting of COVID-19 spread in India. SN Comput. Sci. 2, 416 (2021)
7. Trull, O., García-Díaz, J.C., Troncoso, A.: One-day-ahead electricity demand forecasting in holidays using discrete-interval moving seasonalities. Energy 231, 1–2 (2021)
8. Machine Learning Pills: Step-by-Step Guide to Time Series Forecasting with SARIMA Models (2023)
9. Phumchusri, N., Suwatanapongched, P.: Forecasting hotel daily room demand with transformed data using time series methods. J. Revenue Pricing Manag. 22 (2023)
10. Author, F.: Contribution title. In: 9th International Proceedings on Proceedings, pp. 1–2. Publisher, Location (2010)
11. Tebaldi, L., Bigliardi, B., Filippelli, S., Bottani, E.: EOI or EOQ? A simulation study for the inventory management of a company operating in the railway sector (2023)
12. Caldwell: Economic orden wuantity (EOQ) defined (2021)
13. Mecalux: Stock de seguridad: ¿Qué es y cómo actualizarlo? (2019)
14. Nuzul, F., Sukarno, I.: Safety stock analysis of ship fuel in shipping company (Case study: while oil ship PT. Pertamina Persero) (2020)

15. LNCS. http://www.springer.com/lncs. Accessed 20 June 2023
16. Phumchusri, N., Suwatanapongched, P.: Forecasting hotel daily room demand with transformed data using time series methods. J. Revenue Pricing Manag. (2021)
17. Trull, O., García, J., Troncoso, A.: One-day-ahead electricity demand forecasting. In holidays using discrete-interval moving seasonalities. Energy, 1 (2021)

Optimizing Low-Carbon Job Shop Scheduling in Green Manufacturing with the Improved NSGAII Algorithm

Hongxu Liu[1] , Ping Ting Zhuang[2] , Jinsong Zhang[3] , and Kanxin Hu[1]([✉])

[1] National University of Singapore, Singapore 119077, Singapore
kxhu125@gmail.com
[2] Fuzhou University, Fuzhou 350108, China
[3] Harbin Institute of Technology, Harbin 150001, China

Abstract. In recent times, there has been a growing momentum towards green manufacturing in the production industry, with low-carbon manufacturing emerging as an essential component of this eco-friendly approach. Numerous factories face challenges in transitioning to greener equipment and low-carbon technologies due to constraints such as limited technological capabilities, insufficient investments, outdated management practices, and inadequate human resources. Among various solutions, optimizing job shop management stands out, offering a pathway that requires less investment, lower risk, and lower uncertainty with an enhanced adaptability. This study presents an innovative approach for factories to reduce carbon emissions without compromising production efficiency. This paves the way for factories to achieve green manufacturing and carbon neutrality. First, we establish a low-carbon FJSP (Flexible Job-Shop Scheduling Problem) mathematical model. Then, we employ the enhanced Non-dominated Sorting Genetic Algorithm II to resolve the scheduling problem for the specific factories. With enumerated arithmetic, we finally determine the weight ratio of the low-carbon FJSP model, balancing carbon dioxide reduction and productivity maintenance. Our findings indicate that the integration of carbon emissions into job shop scheduling not only reduces carbon emissions, but also ensures sustained productivity. This research illustrates that the reduction of carbon emissions does not necessarily lead to the decline of productivity in small, medium, and large-scale production. It provides a novel perspective for factories of all sizes to address challenges in green and low-carbon manufacturing.

Keywords: Green manufacturing · Low-carbon manufacturing · Job-shop scheduling · NSGAII · Productivity

1 Introduction

China stands as the predominant energy consumer globally, with its carbon emissions representing roughly 30% of the worldwide total [1]. Notably, the manufacturing sector in China is the principal contributor to these emissions, accounting for 56.1% of the nation's total energy consumption in 2022 [2]. Characterizing the country's manufacturing landscape are factories that are widely distributed, often operate on a smaller scale, and exhibit a lower industrial concentration [3, 4]. These facilities tend to have production processes marked by high input, significant energy consumption, elevated carbon emissions, and relatively low efficiency [5]. China's pledge to attain carbon neutrality by 2060 [6] underscores the pressing need for transformative strategies, among which the adoption of green manufacturing emerges as pivotal [7]. To foster this green transition, the energy-saving and emissions reduction (ESER) strategy has been identified as a cardinal measure [8]. In context of global environmental concerns, low-carbon manufacturing is steadily gaining prominence in the production sector. However, certain factories remain tethered to antiquated manufacturing paradigms, prioritizing productivity and efficiency enhancement as their primary objectives [9]. While there is a growing acknowledgment of the significance of low-carbon manufacturing among various manufacturers, several are reluctant to adapt, primarily due to constraints such as limited technological expertise, financial shortcomings, outmoded management systems, and inadequate human resources [9, 10]. Moreover, there's an underlying apprehension that curtailing carbon emissions might inadvertently compromise productivity, thereby elevating product costs [11]. This potential price escalation could jeopardize the international competitiveness of their offerings [12]. Interestingly, even among factories equipped with the requisite resources for low-carbon manufacturing, there exists a pervasive skepticism regarding the effective execution of carbon-reduction initiatives [13].

Currently, manufacturers predominantly employ three strategies to curtail carbon emissions in their operations: (1) Equipment modernization, encompassing both the transition to green machinery and investment in remanufactured devices. (2) Integration of low-carbon technologies, prominently featuring the adoption of renewable energy sources. (3) Refinement of job shop management practices, which involves enhancing resource utilization rates [11] and making strategic operational adjustments [14]. However, these strategies are not without challenges. Many manufacturing units face financial constraints that hinder their ability to upgrade to environmentally-friendly equipment [15]. Moreover, the disposal of antiquated machinery can lead to significant environmental hazards [5]. While the incorporation of low-carbon technologies offers a sustainable avenue, its adoption is marred by time-intensive processes, considerable uncertainty, substantial investment needs, and a relatively low success ratio [16]. Such endeavors can inadvertently escalate production expenses, leading to increased retail prices and potentially stifling consumer demand [17]. Consequently, to maximize profits, manufacturers might resort to overproduction, which ironically might not yield a tangible reduction in carbon emissions during actual production [18].

In this study, we explore methods to reduce carbon emissions by emphasizing the optimization of job shop management. We introduce a mathematical model addressing the low-carbon flexible job-shop scheduling issue and employ the enhanced NSGAII algorithm to resolve the scheduling problem for one manufacturing factory. Our findings underscore that integrating carbon emissions considerations into job shop scheduling effectively curtails emissions while ensuring sustained productivity.

2 Methodology

2.1 Low Carbon FJSP Mathematical Model

The low carbon FJSP can be described as N jobs are processed on m machines; each job has o_i processes; each process can be completed in optional machines. And the objective function of the model is as follows:

$$\min(f_1 + \alpha \cdot f_2)$$

In the above formula, $f_1 = \min(\max C_{ijk})$ represents the minimum makespan and $f_2 = \min(\sum_i \sum_j \sum_k P_{ijk} T_{ijk} + \sum_k^m P_k T_k)$ represents the minimum carbon emission (Table 1).

The constraints of this model are as follows:

s.t.

$$\sum_{j=1}^{o_i} \sum_{k=1}^{m} M_{ijk} = o_i \tag{1}$$

$$\sum_{k=1}^{m} ST_{i(j=1)k} \geq \sum_{k=1}^{m} C_{ijk} \tag{2}$$

$$\sum_{k=1}^{m} M_{ijk} = 1 \tag{3}$$

$$ST_{ijk} \geq ST_{(i-1)jk} + T_{(i-1)jk} \tag{4}$$

$$ST_{ijk} \geq T_{ijk} + C_{ijk} \tag{5}$$

$$ST_{ijk} \geq 0 \tag{6}$$

$$T_{ijk} \geq 0 \tag{7}$$

The meanings of the above formulas are as follows:

(1) Each job requires to complete o_i processes;
(2) Priority exists between the different processes of the same job;
(3) Each process can only be handled on one machine at one time;
(4) Each machine can only manufacture one process of one job at one time;
 5) All machines can process without any failures and interruptions;
(6) The processing time of O_{ij} on machine k is non-negative;
(7) The start processing time of O_{ij} on machine k is non-negative.

2.2 Non-dominated Sorting Genetic Algorithm II

To solve this multi-objective Flexible Job-Shop Scheduling Problem (FJSP), we employ the NSGAII algorithm, known for its simplicity and effectiveness. It is one of the most widely used and influential multi-objective genetic algorithms [19]. Detailed algorithm parameters are provided in Sect. 3.

Table 1. Parameters of The Model.

Parameters	Meaning
α	Weighted coefficient
i	Job number, $i = 1, 2, \dots n$
j	Process number, $j = 1, 2, \dots o_i$
k	Machine number, $k = 1, 2, \dots m$
n	Total number of the jobs
m	Total number of the machines
M	Machine set, $M = \{M_1, M_2, \dots M_m\}$
P	Job set, $P = \{P_1, P_2, \dots P_n\}$
o_i	Number of processes for job i
O	Process set, $O = \{O_{ij}\}$
M_{ijk}	Whether the process j of job i is processed on the machine k, if yes = 1, no=0
T_{ijk}	Processing time of process O_{ij} on machine k
T_k	Idle time of machine k
ST_{ijk}	The start processing time of process O_{ij} on machine k
C_{ijk}	The completion processing time of process O_{ij} on machine k
MST_k	The start processing time of machine k
MFT_k	The completion processing time of machine k
ME_k	Total carbon emissions of processing jobs on machine k
P_{ijk}	The carbon emission for per second of processing time for job i, process j and machine k
P_k	The carbon emissions of idle time per second on machine k

3   Experiments and Results

Assume that there are three factories with different size—large, medium and small. The manufacturing process in these three factories can be simplified as a flexible job-shop scheduling problem with $8p$ jobs and $7p$ machines, where the values of p for three factories correspond to 10, 5 and 1 respectively. The specific processing time and the average carbon emissions per second during processing (collecting during the manufacturing processing) are shown in Table 2 and Table 3, where $q = 1, 2, \dots p$. '-' indicates the process cannot be handled on the machine.

Table 2. Processing Time.

Job number	Process number	M_{7q-6}	M_{7q-5}	M_{7q-4}	M_{7q-3}	M_{7q-2}	M_{7q-1}	M_{7q}
P_{8q-7}	O_{11}	–	16	–	16	–	15	–
	O_{12}	17	18	–	14	–	–	16
	O_{13}	20	–	20	–	18	13	–
	O_{14}	20	–	17	–	13	–	19
P_{8q-6}	O_{21}	–	17	–	18	–	14	–
	O_{22}	17	–	18	–	22	–	–
	O_{23}	19	17	–	21	19	17	19
P_{8q-5}	O_{31}	18	–	–	20	–	–	–
	O_{32}	18	–	17	23	–	19	–
	O_{33}	–	22	20	–	–	–	17
P_{8q-4}	O_{41}	–	–	–	–	16	–	18
	O_{42}	–	–	–	25	20	21	–
	O_{41}	–	–	–	–	16	–	18
P_{8q-3}	O_{51}	19	–	18	–	16	–	–
	O_{52}	–	12	9	–	–	10	–
	O_{53}	–	–	–	10	13	–	8
	O_{54}	13	16	–	–	14	–	12
	O_{55}	20	–	–	19	–	22	–
P_{8q-2}	O_{61}	10	9	11	9	–	9	10
	O_{62}	–	–	17	–	11	–	–
	O_{63}	14	–	–	10	–	9	–
	O_{64}	–	13	11	–	–	–	10
P_{8q-1}	O_{71}	10	9	–	7	–	–	–
	O_{72}	–	12	–	–	15	–	13
P_{8q}	O_{81}	–	–	11	10	–	8	–
	O_{82}	15	–	–	–	15	–	12
	O_{83}	–	–	6	–	9	13	–

Table 3. Average Carbon Emissions Per Second.

Job number	Process number	M_{7q-6}	M_{7q-5}	M_{7q-4}	M_{7q-3}	M_{7q-2}	M_{7q-1}	M_{7q}
P_{8q-7}	O_{11}	–	1.7	–	1.3	–	1.1	–
	O_{12}	1.2	2.9	–	2.1	–	–	1.6
	O_{13}	1.9	–	1.8	–	1.4	2	–

(*continued*)

Table 3. (*continued*)

Job number	Process number	M_{7q-6}	M_{7q-5}	M_{7q-4}	M_{7q-3}	M_{7q-2}	M_{7q-1}	M_{7q}
	O_{14}	3	–	1.9	–	2.5	–	2
P_{8q-6}	O_{21}	–	2.6	–	2.1	–	1.8	–
	O_{22}	2.6	–	2.2	–	1.4	–	–
	O_{23}	1.5	1.8	–	2.1	2.4	2.1	2.5
P_{8q-5}	O_{31}	2.2	–	–	1.6	–	–	–
	O_{32}	1.7	–	2	2.2	–	1.9	–
	O_{33}	–	2.1	2	–	–	–	1.4
P_{8q-4}	O_{41}	–	–	–	–	1.7	–	1.6
	O_{42}	–	–	–	1.3	1.5	1.4	–
	O_{41}	2.1	–	2.2	–	2.5	–	–
P_{8q-3}	O_{51}	–	2.6	2.2	–	–	2.1	–
	O_{52}	–	–	–	1.9	1.7	–	2.2
	O_{53}	2.7	2.2	–	–	1.5	–	1.9
	O_{54}	1.6	–	–	2.3	–	2.4	–
	O_{55}	1.8	2.1	1.6	1.5	–	2.2	1.5
P_{8q-2}	O_{61}	–	–	1.8	–	2.1	–	–
	O_{62}	2.3	–	–	2.4	–	2.5	–
	O_{63}	–	2.1	1.9	–	–	–	1.1
	O_{64}	1.9	1.6	–	2.4	–	–	–
P_{8q-1}	O_{71}	–	1.5	–	–	1.7	–	1.6
	O_{72}	–	–	2.1	2.4	–	2.2	–
P_{8q}	O_{81}	1.6	–	–	–	1.4	–	1.7
	O_{82}	–	–	2.5	–	2.2	1.5	–
	O_{83}	–	1.7	–	1.3	–	1.1	–

In this paper, we assume that the carbon emissions of the idle time are one-quarter of the carbon emissions of the processing time, as shown in Table 4 below.

Table 4. Carbon Emissions of The Idle Time.

M1	M2	M3	M4	M5	M6	M7
0.5025	0.5275	0.505	0.49	0.4625	0.4825	0.435

We design an improved NSGAII algorithm with the following parameters: (1) The initial population is 100, with 100 iterations; (2) Assume that the initial population is selected using two rules. One rule is to select the minimum processing time of the machine for each process and each job and the other rule is to select the minimum carbon emissions of the machine for each process and each job. The ratio of initial populations selected by these two rules was 1:1; (3) Encoding: double-layer integer encoding of work processes and machines; (4) Process-based POX crossover with a rate of 0.8; (5) Multipoint variation with a rate of 0.01.

This paper aims to proves that considering carbon emissions factors in job shop scheduling can reduce carbon emissions and maintain productivity. The objective function of this model is min $(f_1 + \alpha \cdot f_2)$. When solving this FJSP with the improved NSGAII algorithm, we need to determine the value of α, since different α result in different Pareto fronts. If $\alpha = 0$, the objective function is min(f_1) and the low carbon FJSP mathematical model becomes FJSP mathematical model. Assume that the Pareto set corresponding to $\alpha = 0$ is set A; the Pareto set corresponding to $\alpha = X$ is set B. Therefore, our objective turns to prove that there exists a set B outperforms set A. And we also aim to determine the value of X. Through the enumerate arithmetic, we obtain corresponding Pareto fronts with different α. To simplify the comparison of the Pareto fronts, we connect the points on the Pareto front respectively. We also select five Pareto fronts and combine them into the following line graph (when $p = 1$).

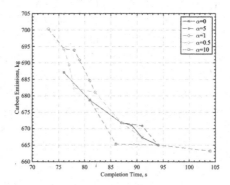

Fig. 1. Pareto Fronts with Different α ($p = 1$)

Our objective is to have less carbon emissions and maintain productivity in manufacturing. If set B exists, for any completion time in Fig. 1, the carbon emissions of set B must be less than or equal to set A at the same completion time. Hence, we eliminate the set with more carbon emissions than set A at the same completion time and obtain the figure below.

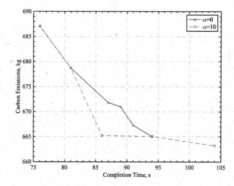

Fig. 2. The Eliminated Pareto Front ($p = 1$)

In Fig. 2, we can observe that the Pareto set of $\alpha = 10$ has lower or same carbon emissions compared with set A. Therefore, set B exists and one of the solutions of X is 10 in this multi-objective FJSP. We prove that considering carbon emissions in FJSP can reduce carbon emissions and maintain productivity in small-scale production (when $p = 1$). The following Fig. 3 shows the job shop scheduling diagram in small-scale production with $\alpha = 10$. Through the above methods, we also demonstrated that reducing carbon emissions does not necessarily bring down the productivity in medium and large-scale manufacturing, as shown in Figs. 4 and 5.

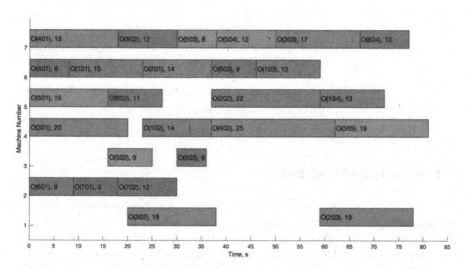

Fig. 3. Job Shop Scheduling Diagram ($\alpha = 10$)

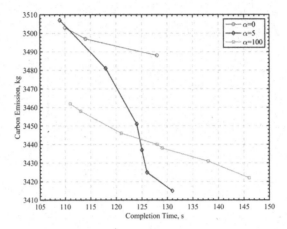

Fig. 4. The Eliminated Pareto Front ($p = 5$)

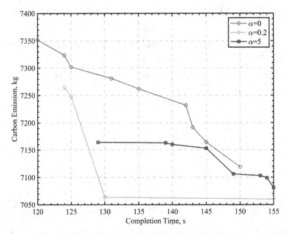

Fig. 5. The Eliminated Pareto Front ($p = 10$)

4   Discussion and Conclusion

This study aims to demonstrate that incorporating the factor of carbon emissions into job shop scheduling not only brings down carbon emissions but also maintains production efficiency, thereby promoting the green manufacturing. Therefore, in addition to considering the traditional scheduling efficiency performance index, the minimum makespan, we introduce the carbon emissions index into the FJSP and establish a dual-objective FJSP model. A NSGAII algorithm, refined with a new population initialization method, is proposed to enhance its efficiency for the optimization process. Subsequently, we ascertain the weight ratio, α, of the low-carbon FJSP model. Our research shows that the Pareto set of $\alpha = 5$ all have approximate or lower carbon emissions than that of $\alpha = 0$ in production of small, medium and large scales. This makes the standardization of carbon reduction methodology possible, paving the way for the establishment of the scheduling models, and job shop scheduling diagrams for factories in all scales.

Our research demonstrates that reducing carbon emissions does not necessarily bring down the productivity in small-scale, medium-scale and large-scale manufacturing. This serves as an up-to-date approach for factories to tackle the challenges of green and low-carbon manufacturing. Currently, our research primarily focuses on optimizing job shop scheduling. Not only is our research beneficial for the decline of carbon emissions in factories, but it facilitates the carbon reduction merging into the entire job shop and supply chain management. Our research goes in line with Chinese strategy of carbon neutrality.

Additionally, there exist several research fields waiting to be dug into in job shop management: (1) The standardization of carbon reduction methodology in job shop scheduling. Whatever the sizes of the factories (small, medium or large) are, we need to radicate a low-carbon FJSP methodology for manufacturing factories of the same production lines; (2) The carbon emissions in the transportation of raw materials and parts; (3) Some practical factors in green manufacturing, such as raw material delay and machine breakdown.

References

1. GE (China) Center for Energy Economics and Research.: GE China Energy Transition White Paper. https://nyjj.xust.edu.cn/info/1090/1715.htm. Accessed 22 Apr 2022
2. National Bureau of Statistics of China 2022: China Statistical Yearbook. www.stats.gov.cn/sj/ndsj/2022/indexeh.htm
3. Wang, X.: The development of low carbon manufacturing industry in Hainan Province. In: MATEC Web of CONFERENCE, vol. 100. EDP Sciences (2017)
4. China Economic Census Leading Group: China Economic Census Yearbook. www.stats.gov.cn/sj/pcsj/jjpc/4jp/indexch.htm. Accessed 2018
5. Cao, H., Du, Y., Chen, Y.: Exploring a new low-carbon development paradigm for China's future manufacturing sectors. J. Sci. Technol. Policy China 2(2), 159–170 (2011)
6. Jia, Z., Lin, B.: How to achieve the first step of the carbon-neutrality 2060 target in China: the coal substitution perspective. Energy 233, 121179 (2021)
7. CICC Research, CICC Global Institute Zipeng. Zhou@ cicc. com. Cn.: Guidebook to Carbon Neutrality in China: Macro and Industry Trends under New Constraints, Springer, Singapore, pp.143–166 (2022)
8. Cai, W., Lai, K., Liu, C., et al.: Promoting sustainability of manufacturing industry through the lean energy-saving and emission-reduction strategy. Sci. Total. Environ. 665, 23–32 (2019)
9. Zhao, Z., Hou, L.: How china respond to the carbon emissions—challenges, commitment and action. J. Low Carbon Econ. 2 (2013)
10. Kang, K., et al.: Evolutionary game theoretic analysis on low-carbon strategy for supply chain enterprises. J. Clean. Prod. 230, 981–994 (2019)
11. Du, Y., et al.: Life cycle oriented low-carbon operation models of machinery manufacturing industry. J. Clean. Prod. 91, 145–157 (2015)
12. Lee, H.: Is carbon neutrality feasible for Korean manufacturing firms? The CO2 emissions performance of the Metafrontier Malmquist–Luenberger index. J. Environ. Manag. 297, 113235 (2021)

13. Fernando, Y., Hor, W.L.: Impacts of energy management practices on energy efficiency and carbon emissions reduction: a survey of Malaysian manufacturing firms. Resour. Conserv. Recycl. **126**, 62–73 (2017)
14. Benjaafar, S., Li, Y., Daskin, M.: Carbon footprint and the management of supply chains: insights from simple models. IEEE Trans. Autom. Sci. Eng. **10** (1), 99–116 (2012)
15. Li, F., et al.: Can low-carbon technological innovation truly improve enterprise performance? The case of Chinese manufacturing companies. J. Clean. Prod. **293**, 125949 (2021)
16. Zhao, S., Bi, K.: A study on radical innovation mechanism of low-carbon technology in manufacturing. In: 2016 International Conference on Management Science and Engineering (ICMSE) (2016). https://doi.org/10.1109/ICMSE.2016.8365456
17. Luo, Z., Chen, X., Wang, X.: The role of co-opetition in low carbon manufacturing. Eur. J. Oper. Res. **253** (2), 392–403 (2016)
18. Zhao, Y.C., et al.: Joint decisions on low-carbon investment and production quantity in make-to-stock manufacturing. In: Proceedings of the World Congress on Engineering, vol. 2 (2016)
19. Xu, L., et al.: Improved adaptive non-dominated sorting genetic algorithm with elite strategy for solving multi-objective flexible job-shop scheduling problem. IEEE Access **9**, 106352–106362 (2021)

Proffer for the Implementation of the SLP Method and Automation in a Poultry Plant for Process Optimization

Steve Mansilla, Jeremy Saravia$^{(\boxtimes)}$, Nicole Cruz, and Javier Romero

Universidad Continental, Huancayo, Perú

{71222888,75787702,71051947,jromerom}@continental.edu.pe

Abstract. This research analyzes the distribution of a poultry plant, with defined processes and established areas. However, it has been proven that the poultry production process can optimize its processes by redistributing it, applying the SLP (Systematic Layout Planning) methodology. In addition, the purpose is to take the process to a second level of improvement with process automation. Initially, the plant plans were analyzing in detail, these allowed us to visualize which areas should be improved. Then, the DAP (Process Activity Diagram) was analyzed, where the activities are identified for the plant process. Thereafter, the SLP method is applied, which is essential for the redistribution of the plant. Once a more efficient plant has been obtained, productivity is analyzed using a distance and load matrix. Furthermore, the idea of automation in important processes is shown in a cost and time table. Finally, a DAP is developed that gives validity to this proposal.

Keywords: SLP · automation · plant · poultry · optimization · implementation

1 Introduction

Nowadays, companies increasingly face the challenges of a changing and globalized market, and they have to innovate and apply continuous improvement in their operations, an efficient and effective organizational strategy in achieving objectives is to offer good quality products in order to have a competitive advantage and not to be left behind in the market [1].

Within the internal production strategy process that is applied in this market to achieve superior quality and competitive prices, the most important part is plant distribution. Because, essentially, it tends to avoid unnecessary expenses of labor and space, perhaps these are factors of little importance in companies in underdeveloped economies but they are very significant in industries that intend to achieve or have achieved stability [2]. In addition, there will be an impact of automation because in the national and international industrial sector, the lack of several considerations of automation projects is evident, highlighting quality improvements, increases in production, increase in repeatability and stability of manufacturing processes, reduction of physical and repetitive work, obtaining

S.-H. Sheu (Ed.): IEIM 2024, CCIS 2070, pp. 87–99, 2024.
https://doi.org/10.1007/978-3-031-56373-7_8

greater continuity of production on holidays, improvement of the cost-benefit relation-ship, selection of the most viable technical and economic offer in terms of automation technology [3]. In this case, a proposal for the implementation of plant distribution is presented for the best use of the spaces in the poultry plant, improving the distribution of areas, identifying process delays between areas, increasing profits and reducing times of processes.

2 Literature Review

Muther, mentions that by spending time planning the design before installation we can significantly reduce economic losses and time delays. A bad design requires future reor-ganization, which is time-consuming and expensive. For this reason, industries develop different methods and algorithms in order to have adequate planning and design [2]. There are different planning techniques to develop a new design or improve the current design, such as Systematic Design Planning (SLP), Paired Exchange Method (PEM), Graph Based Theory (GBT), Dimensionless Block Diagram (DBD), Total Closeness Classification (TCR), etc. [4]. In this case, the SLP application and implementation of automation will be proposed, because this poultry industry needs to produce variety of products and increase its capacity to compete in the constantly growing market. There-fore, this study focuses on the proposal of a new plant design to increase production capacity and implementation of automation for process optimization in a poultry com-pany. The author describes to us that the Systematic Layout Planning (SLP) method designed by Richard Muther in 1968, allows us to solve plant design problems, uses quantitative data, so we can design an effective layout in order to increase productivity, reduce costs and spaces [5].

2.1 Systematic Layout Planning (SLP)

Characteristics. SLP is a methodological application developed by Muther, which indi-cates eleven steps to follow. With this application it is possible to find many solutions for the layout of a plant [6]. Chien, classified these steps of SLP into four parts which are: data register, product processes, output results and evaluation process [7]. It is impor-tant to mention that he also modified SLP to use this procedure in different shapes and hexagons.

Steps. In this research, the SLP technique has been implemented, which consists of the following six steps [8]:

a) Identify and diagram the relationships between resources, obstacles and the environ-ment based on the flow of material and service.
b) Establish and document space requirements and supplies needed for each resource.
c) Draw a relationship graph to create a graphical representation. The nodes in the graph represent the resources and the edges the relationships. Different types of edges can be used to visualize different relationships.
d) Create space relationship arrangements based on the space requirements of resources, obstacles, and the environment.

e) Evaluate the alternatives considering the importance of the different criteria used in the evaluation.

f) Detail the selected layout and make small adjustments to guarantee the correct operation, so the effective automation can be implemented in the selected areas.

2.2 Automation

Automation corresponds to the need to minimize human intervention in production processes, that is, save labor effort [9]. Therefore, the authors tell that a traditional automation line can increase productivity approximately ten times more than its usual production [4]. Furthermore, professionals confirm that automation is more hygienic, efficient, and environmentally friendly. Which allows having a system of several activities, in the same place, where there are specialized machines that carry out cutting operations and repetitive activities [9].

2.3 SLP and Automation Applied to Industrial Plants

The SLP application and the implementation of automation provide different benefits for the poultry industry such as: Increased efficiency and quality, this automation can do repetitive activities., which maintain industry requirements more effectively, reducing production times and operating costs. Furthermore, through the use of SLP, spaces are optimized so that the available space in the plant is more utilized. In industry, automation can also reduce the human production errors, improving food safety and product quality.

3   Methodology

Initially, the plant plans were analyzing in detail, considering the distribution of the areas, between main and other services. This with the objective of understanding the operation of the company. Then, using a DAP (Process Activity Diagram), the existing activities of the production chain are analyzed, and it evaluate their operational efficiency. Subsequently, it is identified which activities are the most important, between critical and non-critical. This is important to prioritize them when is time to redistribute the plant, and also allows them to know if they are suitable for the implementation of automation. Once the evaluation is completed, the SLP method is applied; it has a series of steps, which were detailed previously. This application allows the redistribution of the plant, with a better distribution of areas and optimization of activities. In order to verify if this methodology worked and was useful, an improved DAP of the new proposal is carried out. Which achieves the objective of reduction the unnecessary activities, as well as downtime. Finally, within the objectives of this research, there is the proposal of automation in certain activities of the company, which can be more efficient when implemented with new machinery, bringing with it benefits such as increased productivity and reduction of human errors. The following Fig. 1 shows the steps of the methodology of this research, presented in a flow chart.

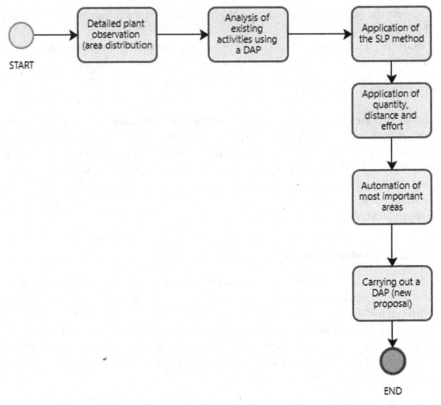

Fig. 1. Methodology flow chart. Own elaboration.

4 Results

4.1 DAP Identification

The DAP, also known as the Process Activity Diagram, is a detailed chart, usually centered on a product component or worker, that represents a variety of elements such as actions, reviews, movements, pauses, storage, schedules, distances, materials, and transportation methods, among others. This diagram enables a detailed analysis of the process [10]. Therefore, we can visualize the company's current DAP below (see Fig. 2).

(DAP)				Operator/material/equipment						
Diagram Nº: Sheet Nº:				Resumen						
				Activity		Current	Proposed		Economy	
Object:				Operation		11				
				Transport		12				
Activity:				Standby		1				
In-plant poultry processing				Inspection		1				
				Storage		2				
Method: Current				Distance (m)						
Location:				Time (min-male)						
Operator: Card Number:				Cost: Labor						
Composed by: Date:				Material						
Approved by: Date:				Total						

Description	Dist. (m)	Time (min)	◯	⇨	D	□	▽	Observation
Receipt of chicken jabas	6	15						Approximately 62 jabas with 12 chickens each
Transfer of jabas to the scale	0	6						With 6 operators
Weighing of jabas	0	48						
Transportation of jabas to the settling area	4	6						For broiler selection and quality control
Chicken Laying	0	30						
Transfer to the stunner	4	6						
Chicken stunning	0	18						It is carried out by means of an electrical stunning machine.
Chicken slaughtering	0	22						Chickens are slaughtered manually by 4 operators.
Transfer to bleeding table	2	6						
Waiting for chick bleeding	0	18						
Transfer to the blanching kettle	4	4						
Chicken scalding	0	32						First scalding
Transfer to the peeling table	4	6						
Chicken peeling		20						
Final peeling inspection	0	22						Manual verification
Transfer to the blanching kettle	4	6						
Chicken scalding	0	18						Second blanching (disinfection)
Transfer to cooling vat	3	0						
Chickens cooling	0	25						
Transfer to cutting table	2	15						
Cutting of limbs	0	30						Manual cutting
Transfer to evisceration table	2	15						
Gutting of chickens	0	40						Manual gutting
Transfer of peeled chickens to vats	2	6						
Washing and disinfection of chickens	0	40						
Transfer to cold storage in jabas	8	6						Capacity of 15 peeled chickens
Final product storage	0	15						For subsequent packaging and distribution
Total	45	475	11	12	1	1	2	

Fig. 2. This DAP (Process Activity Diagram) is the current version managed by the company. It identifies unnecessary transfers and very traditional operations that do not contribute to productivity.

4.2 SLP Application

The main part of plant layout planning is design planning, which focuses mainly on the allocation and layout of functional areas and machines with the objective of optimizing the material flow [11]. Therefore, the application of SLP in the observed plant is presented below (see Fig. 3).

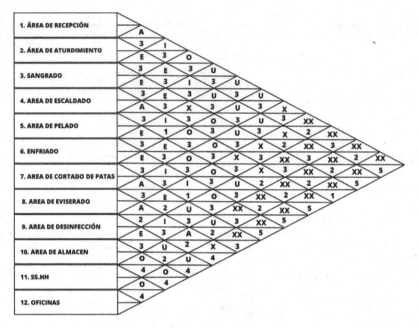

Fig. 3. Relational path diagram, belonging to the SLP methodology for the redesign of a poultry plant.

4.3 Distance, Load and Effort Matrix

Based on the analysis of the sequence of operations, evaluating the quantities to be transported and the distances to travel, the effort that these movements represent is evaluated, giving us this evaluation of the work an index to measure its productivity and propose an improvement proposal [12].

Quantity Matrix. The quantity matrix is fundamental in industrial plant design because it provides a quantitative representation of the necessary resources and their location. We can see in the following Fig. 4, the current Quantity Matrix of the plant under observation.

	Reception of broiler litter	Weighing of bins	Broiler Laying	Chicken stunning	Reception of broiler litter	Chick slaughtering	Waiting for chicken bleeding	Scalding of chickens	Chick peeling	Final peeling inspection	Chickens scalding	Cooling of chickens	Cutting of limbs	Gutting of chickens	Washing and sanitizing chickens
Reception of broiler litter	■	1500													
Weighing of bins		■	1500												
Broiler Laying			■	1500											
Chicken stunning				■	1500										
Chick slaughtering					■	1500									
Waiting for chicken bleeding						■	1500								
Scalding of chickens							■	1500							
Chick peeling								■	1500						
Final peeling inspection									■	1500					
Chickens scalding										■	1500				
Cooling of chickens											■	1500			
Cutting of limbs												■	1500		
Gutting of chickens													■	1500	
Washing and sanitizing chickens														■	15000
Final product storage															■

Fig. 4. The production that the poultry company has in all areas, a quantity of 15,000 chickens monthly, can be identified with the quantity matrix.

Distance Matrix. The distance matrix is an important tool in plant design because it helps engineers and planners make informed decisions about the layout of equipment, work areas, and material flows to achieve an efficient, safe, and profitable plant. By evaluating and minimizing distances between key points, significant benefits can be realized in terms of operational efficiency and productivity. This is represented in the following figure (see Fig. 5).

Effort Matrix. The effort matrix is an important management tool in plant design that helps companies understand and manage the distribution of workloads and resources in their facilities. This facilitates decision making, efficiency optimization and resource planning to ensure efficient and profitable plant operation. Therefore, the effort matrix identified in the company is shown in the following Fig. 6.

	Reception of broiler litter	Weighing of bins	Broiler Laying	Chicken stunning	Reception of broiler litter	Chick slaughtering	Waiting for chicken bleeding	Scalding of chickens	Chick peeling	Final peeling inspection	Chickens scalding	Cooling of chickens	Cutting of limbs	Gutting of chickens	Washing and sanitizing chickens
Reception of broiler litter	■	6													
Weighing of bins		■	4												
Broiler Laying			■	4											
Chicken stunning				■	0										
Chick slaughtering					■	2									
Waiting for chicken bleeding						■	4								
Scalding of chickens							■	4							
Chick peeling								■	0						
Final peeling inspection									■	4					
Chickens scalding										■	3				
Cooling of chickens											■	2			
Cutting of limbs												■	2		
Gutting of chickens													■	2	
Washing and sanitizing chickens														■	8
Final product storage															■

Fig. 5. In the following figure we can identify the distance matrix between areas of the poultry company.

4.4 Distance, Load and Effort Proposed Matrix

The matrix analysis technique identifies critical points in the plant layout and allows the planner to concentrate his effort on the points that offer the greatest probability of introducing an improvement. However, it is not a means of determining the optimal arrangement, but rather of evaluating different possible arrangements on a quantitative and comparative basis. For this reason, it constitutes a valuable technique [12].

Proposed Quantity Matrix. The proposed quantity matrix demonstrated a notable increase in products, being fundamental in the design of an industrial plant by providing a quantitative representation of the necessary resources and their location in the current diagram. In the following Fig. 7, we can see the proposed quantity matrix for the redistribution of the plant under investigation.

Proposed Distance Matrix. The proposed distance matrix minimized the distances between key points, significant benefits can be obtained in terms of operational efficiency and productivity. We can visualize the aforementioned in the following Fig. 8.

	Reception of broiler litter	Weighing of bins	Broiler Laying	Chicken stunning	Reception of broiler litter	Chick slaughtering	Waiting for chicken bleeding	Scalding of chickens	Chick peeling	Final peeling inspection	Chickens scalding	Cooling of chickens	Cutting of limbs	Gutting of chickens	Washing and sanitizing chickens	TOTAL
Reception of broiler litter	■	90000														90000
Weighing of bins		■	60000													60000
Broiler Laying			■	60000												60000
Chicken stunning				■	0											0
Chick slaughtering					■	30000										3000
Waiting for chicken bleeding						■	60000									6000
Scalding of chickens							■	60000								6000
Chick peeling								■	0							0
Final peeling inspection									■	60000						60000
Chickens scalding										■	45000					45000
Cooling of chickens											■	30000				30000
Cutting of limbs												■	30000			30000
Gutting of chickens													■	3000		30000
Washing and sanitizing chickens														■	120000	12000
Final product storage															■	0
15000/615000=0.0244																615000

Fig. 6. In the following figure we can identify the effort of each area and at the same time it is the result of the quantity produced between the resource having the productivity.

	Reception of broiler litter	Weighing of bins	Broiler Laying	Chicken stunning	Reception of broiler litter	Chick slaughtering	Waiting for chicken bleeding	Scalding of chickens	Chick peeling	Cooling of chickens	Cutting of limbs	Gutting of chickens	Washing and sanitizing chickens	Cooling of chickens
Reception of broiler litter	■	37500												
Weighing of bins		■	37500											
Broiler Laying			■	37500										
Chicken stunning				■	37500									
Chick slaughtering					■	37500								
Waiting for chicken bleeding						■	37500							
Scalding of chickens							■	37500						
Chick peeling								■	37500					
Final peeling inspection									■	37500				
Cutting of limbs										■	37500			
Gutting of chickens											■	37500		
Washing and sanitizing chickens												■	37500	
Final product storage													■	37500
Cutting of limbs														■

Fig. 7. The proposed quantity matrix can be identified for the production that the poultry company has in all areas with a quantity of 37,500 chickens monthly.

	Reception of broiler litter	Weighing of bins	Broiler Laying	Chicken stunning	Reception of broiler litter	Chick slaughtering	Waiting for chicken bleeding	Scalding of chickens	Chick peeling	Cooling of chickens	Cutting of limbs	Gutting of chickens	Washing and sanitizing chickens	Cooling of chickens
Reception of broiler litter		6												
Weighing of bins			0											
Broiler Laying				4										
Chicken stunning					0									
Chick slaughtering						2								
Waiting for chicken bleeding							4							
Scalding of chickens								4						
Chick peeling									0					
Final peeling inspection										3				
Cutting of limbs											0			
Gutting of chickens												0		
Washing and sanitizing chickens													2	
Final product storage														8
Cutting of limbs														

Fig. 8. In the figure we can identify the distance matrix proposed between the areas of the poultry company.

Proposed Effort Matrix. The proposed effort matrix helps companies understand and manage the distribution of workloads and resources in their facilities. This will facilitate informed decision making, efficiency optimization and resource planning to ensure efficient and profitable plant operation.

This is demonstrated in the following Fig. 9 shown below.

	Reception of broiler litter	Weighing of bins	Broiler Laying	Chicken stunning	Reception of broiler litter	Chick slaughtering	Waiting for chicken bleeding	Scalding of chickens	Chick peeling	Cooling of chickens	Cutting of limbs	Gutting of chickens	Washing and sanitizing chickens	Cooling of chickens	total
Reception of broiler litter		225000													225000
Weighing of bins			0												0
Broiler Laying				150000											150000
Chicken stunning					0										0
Chick slaughtering						75000									75000
Waiting for chicken bleeding							150000								150000
Scalding of chickens								150000							150000
Chick peeling									0						0
Final peeling inspection										112500					11250
Cutting of limbs											0				0
Gutting of chickens												0			0
Washing and sanitizing chickens													75000		75000
Final product storage														300000	30000
Cutting of limbs															0
375000/61=0.0303															1237500

Fig. 9. In the figure we can identify the proposed effort matrix of each area and at the same time it is the result of the quantity produced between the resource having the productivity.

4.5 Proposal DAP

The DAP proposal (Process Activity Diagram) is presented below in Fig. 10, which reflects the changes generated by the SLP application and automation together, which bring improvements in distances and times.

(DAP)				Operator/material/equipment			
Diagram Nº: Sheet Nº:					Resumen		
				Activity	Current	Proposed	Economy
Object:				Operation	11		
				Transport	12		
Activity:				Standby	1		
In-plant poultry processing				Inspection	1		
				Storage	2		
Method: Current				Distance (m)			
Location:				Time			
				(min-male)			
Operator: Card Number:				Cost:			
				Labor			
Composed by: Date:				Material			
Approved by: Date:				Total			

Description	Dist. (m)	Time (min)	○	⇨	D	▢	▽	Observation
Receipt of chicken jabas	6	15						Approximately 62 jabas with 12 chickens each
Transfer of jabas to the scale	0	6						With 6 operators
Weighing of jabas	0	48						
Transportation of jabas to the settling area	4	6						For broiler selection and quality control
Chicken Laying	0	30						
Transfer to the stunner	4	6						
Chicken stunning	0	18						It is carried out by means of an electrical stunning machine.
Chicken slaughtering	0	22						Chickens are slaughtered manually by 4 operators.
Transfer to bleeding table	2	6						
Waiting for chick bleeding	0	18						
Transfer to the blanching kettle	4	4						
Chicken scalding	0	32						First scalding
Transfer to the peeling table	4	6						
Chicken peeling		20						
Final peeling inspection	0	22						Manual verification
Transfer to the blanching kettle	4	6						
Chicken scalding	0	18						Second blanching (disinfection)
Transfer to cooling vat	3	0						
Chickens cooling	0	25						
Transfer to cutting table	2	15						
Cutting of limbs	0	30						Manual cutting
Transfer to evisceration table	2	15						
Gutting of chickens	0	40						Manual gutting
Transfer of peeled chickens to vats	2	6						
Washing and disinfection of chickens	0	40						
Transfer to cold storage in jabas	8	6						Capacity of 15 peeled chickens
Final product storage	0	15						For subsequent packaging and distribution
Total	45	475	11	12	1	1	2	

Fig. 10. DAP (Process Activity Diagram) Proposed version.

5 Discussion

A case of SLP application is the research called "Design performance indicators and systematic planning", where the objectives are to carry out plant layout redesign planning applying this methodology, with this you can reduce transfer times and routes. The final results of this research were to minimize the area of the production machinery and then place it in another part of the plant, with this more space was reduced that allowed the other areas to be organized. And have less displacement of operators. Finally, this application increased staff productivity and improved utilization of the production area [13]. Comparing the initial situation with the current one, productivity will be defined using the following expression: Productivity = Production/Resources Where "resources" will be represented by the effort or work generated by the transfer of material from one place to another [12]. For this case, the variation in productivity will be determined in order to evaluate the proposed alternative having an improvement of 24.24% in productivity.

6 Conclusion

In conclusion the application of SLP together with automation has the potential to significantly transform operations in a variety of industries and sectors. It offers benefits such as improved efficiency, error reduction, resource optimization and increased flexibility, which can lead to increased competitiveness and efficiency in a wide range of applications.

In addition, it is also concluded that this study demonstrates that a good plant layout and automation of critical processes improves the productivity of any company, regardless of industry or size. In this way, it avoids business dwarfism in third world countries and improves the quality of products and services.

7 Recommendation

It is recommended to follow the detailed guidelines on how to carry out the implementation of SLP and automation together, if you want to apply this proposal in any industry. Additionally, a step-by-step implementation plan is suggested, highlighting the resources required and approximate timelines.

References

1. Chagua Magro, D.N.: Propuesta de rediseño de planta para incrementar la productividad de la producción de pollos de engorde en la empresa San Fernando del distrito de Lima (2021)
2. Muther, R.: Distribución en planta. Hispano Europea, Barcelona (1981)
3. Cordoba Nieto, E.: Manufactura y automatización (2006)
4. Alvseike, O., et al.: Meat inspection and hygiene in a Meat Factory Cell – an alternative concept. Food Control (2018)
5. Muther, R., Hales, L.: Systematic Layout Planning, 4th edn., vol. 4. Management and Industrial Research Publications (2015)
6. Muther, R.: Systematic Layout Planning. Cahners Books, Bóston (1963)
7. Chien, T.-K.: An empirical study of facility layout using a modified SLP procedure, vol. 15 (2004)
8. Elahi, B.: Manufacturing plant layout improvement: case study of a high-temperature heat treatment tooling manufacturer in Northeast Indiana. Procedia Manuf. 53 (2021)
9. de Medeiros Esper, I., From, P.J., Mason, A.: Robotisation and intelligent systems in abattoirs (2021)
10. Bocangel, G., et al.: Ingeniería Industrial - Ingeniería de Métodos I (2020)

11. Klar, M., Langlotz, P., Aurich, J.: A framework for automated multiobjective factory layout planning using reinforcement learning (2022)
12. Días, B., Jarufe, B., Noriega, M.T.: Disposición de planta, Lima (2007)
13. Flessas, M., Rizzardi, V., Tortorella, G.: Indicadores de desempeño del diseño y planificación sistemática: un estudio de caso en un restaurante del sur de Brasil (2015)

Optimization of the Stacking Process of Wire Mesh Coils in Industrial Processors

Renzo Andree Rojas Benito⬥, Stefanny Pamela Inocente Hurtado,
Pamela Stefany Carrion Miguel(✉) ⬥, and Guillermo Anibal Bayona Carazas

Department of Industrial Engineering, Universidad Continental, Huancayo, Peru
74074875@continental.edu.pe

Abstract. This study proposes to optimise the post-production and storage process of electro-welded mesh rolls in companies in the city of Lima, for which this study has taken into account the entire production line and production of electro-welded mesh up to its storage, where the bottlenecks encountered occur when the final product is removed from the production line and taken to the warehouse. The forklift operator has to come constantly from time to time to lift the rolls two by two onto a rack and then take them to the warehouse with a total of 15 rolls, and this process is repeated every day. In the warehouse, the rolls are stacked two by two, all of which takes time and delays productivity. That is why the aim of this study is to analyse production, post-production and storage and to optimise to improve time, production and resource management. All this is achieved by using mathematical calculations, analysis software and finally an automatic machine is proposed, so that when the roll comes out of the winding machine, it is automatically stacked in the rack, once the 15 rolls are stacked, the conveyor belt moves the entire rack and only the forklift moves to the warehouse, being more productive since the racks will also be stackable. The results obtained in the analysis show that by implementing an automatic post-production and storage machine, the transfer and stacking times decrease significantly up to 76.76% and the operational efficiency of the machine improves up to 79.35%, which benefits greatly and in the long term the entire production of this industrial line.

Keywords: Hardware · Software · Design · Optimization

1 Introduction

Nowadays, at an international level, companies have a constant objective, one of the most important interests for any company is productivity, so that the higher the performance, the higher the profits and the greater the growth. For it, they determine factors that limit and favour the efficiency with the purpose of taking measures and strategies to direct the production adequately [1], In the production line of any company, frequent problems exist that obstruct the fulfilment of the proposed goal, one of the most well-known are the "bottlenecks", characterized for being the causative one in diminishing or affecting the production process [2], leading to a considerable decline in line capacity. Therefore, in order to reduce the problem and increase production, many companies explore new technologies [3], new organisational structural models and time management, so that everything is favourable and costs are reduced [4].

Many companies bet on the development and exponential growth of their products, today we see a lot of industrial companies with countless products and their derivatives, where they are processed and marketed. One of the most common industrial products is wire, whose product is crucial for its diverse applications in various sectors such as construction, agriculture, engineering, among others [5]. However, there are times when a production imbalance is generated, as there are tight and limiting production practices [6], due to different factors, such as: environmental conditions, manufacturing processes, type of machinery, lack of innovation, constant machinery failures, among others [7]. Although there are currently technologies to improve the production of wire (Pd - 5Ni alloy), new intelligent predictive models to improve productivity, this production imbalance is still evident [8].

Thus, in order for the wire to continue to grow constantly, it has diversified from its processing and manufacture. This raw material in 'first instance begins with a treatment that is performed to the wire, this process is the pickling, allowing to remove impurities and leave it in suitable conditions for the drawing process [9]. Later in the drawing process, the wire is processed and comes out with a thinner and smaller diameter than the one it entered. The wire is then placed in a spider winder, which is in a suitable condition for the drawing process [9]. Finally, the wire exits the spider and undergoes a zinc plating treatment [10]. From this process the wire is further diversified into different products, among the most commercial ones are rolls of wire mesh for the construction sector, livestock, gardening and others. Of all these products, the one that is produced in the greatest quantity is electrowelded wire mesh, which is mainly used in the construction sector.

Electrowelded mesh is very important in the construction sector, as time goes by, countries continue to develop and with it comes the improvement of works, making use of new technologies, new raw materials, more consistent than the traditional. This product is mainly used for floors, mezzanines, tunnel linings, among others, giving greater support to columns and foundations [11]. Also, its production continues to grow due to the benefits such as easy installation, economic material, thermal performance, among others, and also due to the demand for projects and works in progress in developing countries [12, 13]. For the processing of electrowelded mesh, a process of electrowelding of the wire is carried out after drawing, giving as a final product a mesh [14].

In China, they present smart manufacturing technologies, greatly benefiting their labour productivity in manufacturing companies [15]. Also in Canada, they show the automation of their factories, so that they integrate production interface with Industry 4.0, allowing factories to be more innovative and with higher productivity [16]. Another study from China presents the impact on the adoption of industrial robots, which aims to promote the impact of technology with the transformation of economic models in relation to sustainable development, so that production is cleaner and helps the environment [17].

In Peru, the development of construction continues to grow, and so does the demand for electrowelded mesh. At present, in the city of Lima, there are companies that process wire and produce electrowelded mesh, many of these companies produce the final product using traditional machines, While it is true that their productivity is enormous, it is still true that the distribution and processing of the products takes time and is a limitation, all of this could be improved if cutting-edge technology is used, which would not only increase production, but would also bring great benefits and development for the country [18].

Based on the detailed analysis of the problem outlined above, the urgency to address the deficiencies in the production and storage process of wire mesh coils in companies in Lima is evident. The direct connection between productivity, the processing time of electrowelded wire mesh and the consequent complications in post-production and storage underlines the need for strategic intervention.

The problems identified, centred on disorder, the prolongation of production time and the consequent inefficient use of labour, become a significant obstacle to the efficient development of industrial activities. This scenario not only affects production capacity in real time, but also results in inefficient management of finished products and their subsequent storage.

In response to these challenges, the present research is proposed as a concerted effort to comprehensively analyse and optimise the post-production and storage process of wire mesh coils. The implementation of mathematical modelling, design and CAD simulation emerge as crucial tools in this endeavour, as they offer the possibility to virtually visualise and evaluate different improvement scenarios.

The optimisation of these processes is not only limited to production efficiency, but also addresses the overall management of the product, from its creation to its final storage. The integration of advanced technologies and innovative approaches aims not only to speed up production, but also to improve the quality of the final product and reduce associated waste.

In summary, this research not only identifies the existing difficulties in the production cycle of wire mesh coils in Lima, but also sets out a clear path towards improvement through the application of advanced techniques. The holistic perspective adopted addresses not only processing efficiency, but also integrated product management, thus opening up a horizon of opportunities for a more competitive and sustainable industry.

2 Methodology

This proposed research has a continuous improvement in the stacking processes of electro-welded mesh rolls as the bottleneck was identified when transferring the finished product to the warehouse. Currently the process is done traditionally, in this study software, mathematical calculations and simulations were used in order to improve the productivity of the production and storage of electro-welded mesh rolls by a high percentage. The process of making electrowelded mesh is shown below.

The detailed manufacturing process of electrowelded wire mesh rolls reveals a series of critical stages that define the quality and efficiency of the final product. Starting with the storage of the wire rod, which is essential as a raw material, the production cycle progresses through crucial phases that not only transform the material but also determine the strength and durability of the finished product.

The initial wire rod pickling stage lays the foundation for a clean wire free of impurities, ensuring the integrity of the product. The wire drawing process then seeks to further

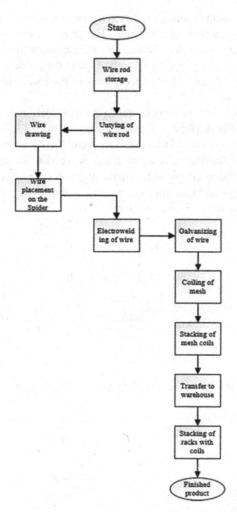

Fig. 1. Flow diagram for the production of electrowelded mesh.

refine the characteristics of the material, ensuring uniformity and desired strength. The introduction of the Spider in this process not only simplifies wire handling, but also plays a crucial role in efficient uncoiling.

Electrowelding, the next step in the production chain, not only joins the wire precisely, but also represents a key milestone in wire mesh manufacturing. The possibility of including the galvanising process prior to electro-welding underlines the versatility of the method, allowing additional protective measures to be applied to the steel and thus improving the corrosion resistance of the final product.

The manual movement of coils to the warehouse reveals a critical facet of the production chain. Identifying the bottleneck in this transfer process underlines the importance of optimising logistics operations to ensure a continuous and efficient flow. Strategies such as the implementation of automated systems or ergonomic improvements can be considered to address this specific challenge. The process culminates with the stacking of the coils on racks in the warehouse, closing the production cycle in an orderly and efficient manner. Figure 1, which visually depicts the procedure, provides a clear snapshot of the complexity and interconnectedness of each stage in the manufacture of electrowelded wire mesh.

The research has been of an experimental type, an evaluation was carried out and the problem was identified in the final process of the production of electrowelded mesh, as when transferring the finished product to the warehouse there is a delay time, this process is currently carried out traditionally using forklifts, so that having rolls on the floor accumulated, there has been a mess, so that in the research simulations and mathematical calculations were used to obtain effective results in improvement to be made in the identified bottleneck as shown in Table 1 and 2.

Table 1. Welded mesh data

Production data collected - Electrowelded mesh 2.50 × 2.40		
Actual production	rollo/sem	470
Planned production	rollo/sem	600
Scheduled operator rest	min/turno	20

Table 2. Labour Productivity Efficiency Rankings

Productivity efficiency	Range
Very Low	10%–40%
Low	41%–60%
Normal	61%–80%
Very Good	81%–90%
Excellent	91%–100%

Productivity efficiency was considered normal and average when it is in the range 61% and 80% of the Table 2, as it can vary from 0 to 100%.

The formula used in the research was shown below as shown in Eq. 1.

$$Productivity = \frac{No. \, of \, units \, produced}{Inputs \, used} \tag{1}$$

Where, productivity is defined as the ratio of the number of units produced to the input used.

In addition, the calculation of output in man-hours is used as shown in Eq. 2:

$$R = \frac{TxN}{V} \tag{2}$$

Where, R is the output in man-hours/unit, T is the Time in duration of the activity, N is the Number of workers involved in the operation and of the same category and V is the volume of work employed.

Also, the current total effectiveness and efficiency with respect to the proposed machine was calculated.

Operational efficiency (production):

Therefore, a total efficiency = 78.3% is obtained.

Labour efficiency:

$$\frac{14*5*4}{(60.50-20)*20} = \frac{280 \text{ min/m}}{810 \text{ min/m}} x100 = 0.3456 \tag{3}$$

Therefore, a total efficiency equal to 34.56% is obtained. Thus, having found these two parameters, the current productivity in the netting production area is calculated. Total productivity is equal to Effectiveness multiplied by Efficiency, i.e. 78.3% × 34.56%, resulting in 27.06%. In order to reach the optimisation of post-production and storage, mathematical calculations were carried out based on simulation data of the stacking process with the proposed machine:

Machine Efficiency (OEE)

$$D = \frac{Productive\ time}{Time\ available} = \frac{1332 \text{ min}}{1440-10} x\ 100 = 93.14\% \tag{4}$$

$$R = \frac{Actual\ production}{NominalCap.\ x\ Active\ Time} \tag{5}$$

$$R = \frac{235\ rollos/d}{(12\ x\ 1332 \text{ min})60 \text{ min}} x\ 100 = 88.21\% \tag{6}$$

$$C = \frac{Good\ parts}{Actual\ productionl} = \frac{235-8\ coils}{235\ coils} x\ 100 = 96.59\% \tag{7}$$

Where the efficiency of the machine is equal to the following:

OEE is equal to the multiplication of D times R times C, i.e. 0.9314 multiplied by 0.8821 and multiplied by 0.9659 giving an equality of 79.35%. Thus, the estimated OEE of the proposed machine is at an "admissible" level which means that it contributes to the improvement of the process. Everything we used helped us to efficiently carry out the analysis of the problems of the identified process, using these mathematical calculations allowed us to clearly see the performance that the workers are executing.

3 Results

The automatic stacking design for the bundling of electrowelded mesh rolls optimises space and time efficiently, as it allows direct access to the mesh rolls. This is due to the integration of software, 3D simulation of a machine as shown in the Fig. 2.

The use of AnyLogic software, as shown in Figs. 3 and 4, has represented a significant advance in the optimisation of the processes linked to the production of netted rolls. This

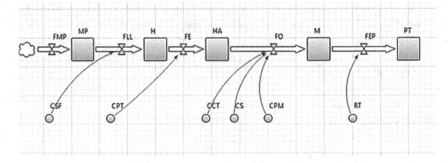

Fig. 2. Forrester Diagram in Anylogic

specialised tool has made it possible to carry out a detailed analysis, delving deeper into the critical aspects of the post-production and storage process. The identification of problems, especially in time and traditional transport, has provided a solid basis for strategic decision-making.

The programme's ability to generate simulations has been instrumental in the design and implementation of improvements in the production process. The effectiveness of the analysis performed through AnyLogic has paved the way for the identification of specific areas requiring attention and adjustments. This predictive capability has made it possible to anticipate potential bottlenecks and propose solutions prior to their implementation in the real production environment.

It is relevant to note that modelling in AnyLogic is not only limited to running simulations, but has also been an integral part of a broader process starting with data analysis. Data collection, verification and validation have been essential steps that have provided the solid basis for decisions made during the implementation of improvements. This data-driven approach ensures that decisions are made in an informed manner and backed by an accurate understanding of the situation.

The model cycle at AnyLogic starts with data analysis, expands into the identification of areas for improvement, and culminates in the implementation of solutions to drive continuous improvement in the production and storage of welded wire mesh rolls. This holistic approach, supported by advanced simulation technology, proves to be an effective strategy for addressing operational challenges and optimising industrial processes.

This design has had a significant impact on the actual time required from cutting the mesh to forklift transfer. The reduction in operating time from 65.5 min to 15.22 min per day represents an extraordinary 76.76% improvement in operational efficiency. This achievement not only translates into time optimization, but also means a considerable increase in production capacity, with the ability to handle 15 racks of 15 rolls each.

The efficiencies generated by the implementation of this design are not only limited to time reduction, but extend to substantial improvements in logistics and storage capacity. With the time saved of 50.28 min, an expansion in rack rotation is projected, which translates into a significant increase in storage capacity. The estimate of being able to handle 11 additional racks suggests a total capacity of 26 racks within the production line, thus optimizing space utilization and improving efficiency in inventory management.

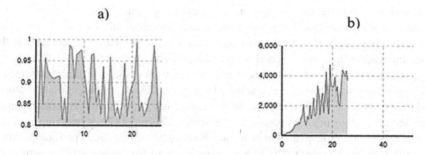

Fig. 3. Plot of variance of a) independent variable and b) Dependent variable.

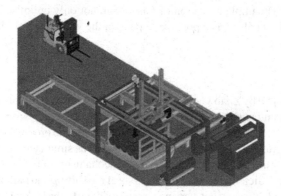

Fig. 4. Proposed machine for post-production and storage improvement.

Furthermore, it is emphasized that the design has been conceived not only to reduce times and increase capacity, but also to improve Overall Equipment Effectiveness (OEE). The formula (8) used to calculate OEE is a key indicator in this design, providing an accurate measure of overall equipment efficiency. The improvement in OEE not only validates the positive impact of the design in terms of availability, quality and perfor-mance, but also establishes a clear framework for ongoing assessment of equipment performance over time.

$$OEE = D \, x \, R \, x \, C = 0.9314 \, x \, 0.8821 \, x \, 0.9659 = 79.35\% \qquad (8)$$

The values for availability (D), quality (C) and performance (R) within the range of 0 to 1 indicate that the Overall Equipment Effectiveness (OEE) is in an "acceptable" category. When calculating the OEE in line with comparable study times, the opera-tional efficiency of the machine is estimated at 79.35%. This result underlines that the performance of the equipment is satisfactory and contributes effectively to the mesh roll stacking process.

In a broader analysis, the implementation of this design not only results in measurable efficiency, but also brings tangible benefits in terms of organisation and safety in the production environment. The demonstrated reduction in operating times is only part of the equation, as a better distribution of space and an optimised transfer order between operators is also observed. This improvement in ergonomics and internal logistics not only adds efficiency to the process, but also decreases the likelihood of accidents in the plant, thus improving working conditions and safety for employees.

In addition to the internal benefits, the implementation of this design presents a valuable opportunity for the sector as a whole. By using efficient equipment that organises products on a production line, it sets a standard of quality and efficiency that can be replicated and adopted by other companies in the sector. This practice not only improves the internal operations of the plant, but also contributes to the advancement and competitiveness of the sector as a whole. The positive assessment of operational efficiency through OEE not only validates the equipment's allowable performance, but also highlights tangible improvements in the organisation, safety and overall efficiency of the web roll stacking process. The implementation of this design not only optimises the company's internal operation, but also has a positive impact on the wider industry landscape.

4 Discussion

In the study developed by Zhao J. and Gao Z. present on an automatic sorting and stacking system for the manufacture of rotary plates, which reveals the cause of the problem identified in the process of sorting and manual palletising in the production line, therefore an automatic integrated device was designed where it was simulated through the Adams kinematics in order to verify the feasibility of the mechanisms of the bars [19]. Although this research shows efficient results in the analysis of the sorting and palletising process, it has not been as productive, so it is very necessary to identify the bottleneck to implement the improvement. In contrast, in the identified process, Anylogic simulation software was used to analyse the problem in order to make decisions on the improvement that was used in the process of manufacturing the electro-welded mesh rolls. In the study of Qin, J. and Wang, J. show the development of a software and hardware framework based on deep learning for a non-contact inspection platform for aggregate classification, this study developed a self-developed test and sampling platform for aggregate recordings quickly presenting results that using the AS Mask RCNN network model achieved 89.13% accuracy in the three experimental situations [20]. Despite the promising results obtained in this study, it is recognised that its efficiency could be limited in the comprehensive analysis needed to drive improvements and carry out the precise implementation in the production process. In order to carry out an effective and meaningful analysis in the field of continuous improvement of production processes, it is essential to have real data from the operating environment. In this context, the proposed study opted for the use of simulations and mathematical calculations to analyse the production process of electrowelded mesh rolls.

The choice of simulations and mathematical calculations allowed a precise identification of the deficiencies in the process, particularly in the segment where the improvements and implementation of the proposed machine were applied. It should be noted that Bolender, S. and Oellerich J. present an innovative system for the stacking of pallet cages by means of a bridge crane - Krass, with the aim of achieving reproducible, automatic and safe stacking without modifying the existing structure [21]. Although this system is semi-automatic, it motivated the specific analysis in the production of electro-welded mesh rolls.

The use of AnyLogic simulation software was instrumental in this analysis, as it allowed the bottleneck in terms of efficiency to be clearly identified. This precise identification of the area for improvement drove the implementation of the proposed automatic machine, thus supporting the philosophy of continuous process improvement. The introduction of this automatic machine is aligned with the objective of optimising the company's productivity, overcoming the limitations of semi-automatic systems and improving the overall efficiency of the production process of electrowelded mesh rolls.

In conclusion, this integrative approach, which combines simulations, mathematical calculations and the implementation of innovative technologies, proves to be an effective strategy to address the complexities of analysing and improving production processes. The adoption of automated solutions, supported by accurate data obtained through simulations, highlights the commitment to continuous improvement and the achievement of higher productivity in the company.

5   Conclusions

The results obtained in this study reveal that the implementation of an automatic machine in the post-production and storage process not only optimizes the final production, but also achieves automated stacking of the electro-welded mesh rolls. The significant 79.35% improvement in operational efficiency clearly indicates the positive impact of this innovation on the overall process performance.

The validation of the machine using the 3D mechanical design simulator (Autodesk Inventor) has contributed to the certification of its efficiency. This approach not only ensures the proper layout of the final products, but also highlights the importance of using advanced tools for efficient resource and space management in the production environment.

The analysis and continuous improvement of the identified bottleneck in the post-production process has been rigorously supported by mathematical calculations. The comparison between the current efficiency and the expected efficiency of the machinery provides a quantitative view of the improvement achieved, showing a time reduction of 76.76%. This concrete figure not only validates the effectiveness of the interventions carried out, but also provides a clear parameter to measure and further improve in the future.

The optimization of processes, as reflected in the significant reduction of times, has a direct impact on improving productivity and efficiency in the post-production phase. This optimization not only translates into internal benefits, but also positions the company to be more competitive in the market by delivering high quality products in more efficient times.

In addition, the process analysis in the AnyLogic software has been an invaluable decision-making tool. This platform has provided a holistic view of operational aspects, enabling the identification of efficient distribution strategies and the assessment of the feasibility of incorporating more state-of-the-art machinery. This proactive approach supports the company's vision towards continuous innovation and adaptation to advanced technologies to remain at the forefront of its sector.

References

1. Rámirez-Ambríz, L., Ojeda-Ruiz, M.Á., Marín-Monroy, E.A., Toribio-Espinobarros, B.E.: Factors that facilitate or limit the development of bivalve mollusk aquaculture in BCS, Mexico: the small-scale producers' perspective. Reg. Stud. Mar. Sci. **66** (2023). https://doi.org/10.1016/j.rsma.2023.103145

2. Javadi, S.M., Sadjadi, S.J., Makui, A.: Identification and fixing bottlenecks of a food manufacturing system using a simulation approach. Sci. Rep. **13**(1) (2023). https://doi.org/10.1038/s41598-023-39025-5

3. Tremblay, F.-A., Durand, A., Morin, M., Marier, P., Gaudreault, J.: Deep reinforcement learning for continuous wood drying production line control. Comput. Ind. **154** (2024). https://doi.org/10.1016/j.compind.2023.104036

4. Zyoud, R.N.: Factors influencing time management skills among nurses in North West Bank, Palestine. BMC Nurs. **22**(1) (2023). https://doi.org/10.1186/s12912-023-01560-x

5. Matias Gonçalves, L., Mendoza-Martinez, C., Alves Rocha, E.P., Coutinho de Paula, E., Cardoso, M.: Solar drying of sludge from a steel-wire-drawing industry. Energies **16**(17) (2023). https://doi.org/10.3390/en16176314

6. Venugopal, V., Saleeshya, P.G.: Productivity improvement through the development of sustainability metrics in wire manufacturing industry. Int. J. Product. Qual. Manag. **39**(1), 1–19 (2023). https://doi.org/10.1504/IJPQM.2023.130891

7. Blivet, C., Larché, J.-F., Israëli, Y., Bussière, P.-O.: Non-Arrhenius behavior: influence of antioxidants on lifetime predictions for materials used in the cable and wire industries. Polym. Degrad. Stab. **201** (2022). https://doi.org/10.1016/j.polymdegradstab.2022.109978

8. Konstantinov, I.L., et al.: Investigation modes for production technology of wire from the Pd5Ni alloy for catchment gauzes of the chemical industry. Int. J. Adv. Manuf. Technol. **121**(11–12), 7229–7246 (2022). https://doi.org/10.1007/s00170-022-09821-w

9. Radionova, L.V., et al.: Mathematical modeling of heating and strain aging of steel during high-speed wire drawing. Metals **12**(9) (2022). https://doi.org/10.3390/met12091472

10. Kustra, P., Wróbel, M., Dymek, S., Milenin, A.: Novel drawing technology for high area reduction manufacturing of ultra-thin brass wires. Arch. Civ. Mech. Eng. **23**(3) (2023). https://doi.org/10.1007/s43452-023-00677-9

11. Massone, L.M., Nazar, F.: Analytical and experimental evaluation of the use of fibers as partial reinforcement in shotcrete for tunnels in Chile. Tunn. Undergr. Space Technol. **77**, 13–25 (2018). https://doi.org/10.1016/j.tust.2018.03.027

12. Zhou, F., Zhou, J., Li, X., Chen, Q., Huai, X.: Enhanced capillary-driven thin film boiling on cost-effective gradient wire meshes for high-heat-flux applications. Exp. Therm. Fluid Sci. **149** (2023). https://doi.org/10.1016/j.expthermflusci.2023.111018

13. Boschi, K., di Prisco, C., Flessati, L.: An innovative design approach for anchored wire meshes. Acta Geotech. **18**(11), 5983–6005 (2023). https://doi.org/10.1007/s11440-023-02046-5

14. Castillo, M.J., María, V.L.C., Marconi, C.D., Svoboda, H.G.: Effect of welding parameters on welded steel wire mesh. Rev. Mater. **23**(2) (2018). https://doi.org/10.1590/S1517-707620180002.0352

15. Zhu, M., Liang, C., Yeung, A.C.L., Zhou, H.: The impact of intelligent manufacturing on labor productivity: an empirical analysis of Chinese listed manufacturing companies. Int. J. Prod. Econ. **267** (2024). https://doi.org/10.1016/j.ijpe.2023.109070

16. Kwok, T.H., Gaasenbeek, T.: A production interface to enable legacy factories for industry 4.0. Eng. Res. Express **5**(4) (2023). https://doi.org/10.1088/2631-8695/acfeca

17. Liang, L., Lu, L., Su, L.: The impact of industrial robot adoption on corporate green innovation in China. Sci. Rep. **13**(1) (2023). https://doi.org/10.1038/s41598-023-46037-8

18. Quiroz, L.G., Maruyama, Y., Zavala, C.: Cyclic behavior of thin RC Peruvian shear walls: full-scale experimental investigation and numerical simulation. Eng. Struct. **52**, 153–167 (2013). https://doi.org/10.1016/j.engstruct.2013.02.033

19. Zhao, J., Gao, Z.-Y., Shen, Y.-S., Song, Y.: Automatic grading and stacking system for rotary veneer manufacturing. Chin. J. Wood Sci. Technol. **35**(5), 66–71 (2021). https://doi.org/10.12326/j.2096-9694.2020179

20. Qin, J., et al.: Deep learning-based software and hardware framework for a noncontact inspection platform for aggregate grading. Meas. J. Int. Meas. Confed. **211** (2023). https://doi.org/10.1016/j.measurement.2023.112634

21. Bolender, S., Oellerich, J., Braun, M., Golder, M., Furmans, K.: System for a reproducible, automatic and safe stacking of pallet cages using an overhead crane – Krass. Logist. J. **2018** (2018). https://doi.org/10.2195/lj_Proc_bolender_de_201811_01

The Effects of Voice Emotions on Users' Willingness to Pay Decision-Making Process of Automated Delivery Robots: An ERP Study

Li Yan, Xie Qiling[(⊠)], and Song Wu

Huaqiao University, Xiamen 350200, China
liyan73@hqu.edu.cn, xql1360804380@outlook.com

Abstract. Automatic delivery robots are increasingly used in various business and daily life scenarios. The advancement of voice interaction technology allows users to engage with these robots naturally. User experience during this interaction is crucial for acceptance and satisfaction. This study aims to use electroencephalogram (EEG) technology to investigate users' cognitive states and emotional responses to different robot voices during the willingness to pay decision-making process. Participants in the experiment rated their willingness to pay based on their interaction with the robot's speech. The findings revealed a significant correlation between speech emotional intensity and willingness to pay. This correlation was supported by variations in EEG signals and there are significant differences in brain activation intensity across different levels of willingness to pay. Identifying this correlation between willingness to pay and EEG signals provides businesses with a novel approach to predict and influence consumer purchasing decisions. It also offers insights for voice design in delivery robots. By understanding the impact of emotional intensity on willingness to pay, businesses can customize robot voices to enhance user experience and increase customer satisfaction.

Keywords: Automatic Delivery Robot · Voice Design · Emotional Speech · User Experience · EEG · WTP

1 Introduction

In the context of global collaborative efforts to implement sustainable transportation solutions [1], the introduction of automated delivery robots is a crucial initiative. The COVID-19 pandemic has led to a surge in public interest and demand for auto-mated delivery robots, as they can provide contactless delivery, which is highly favored under social distancing guidelines. Additionally, with the aging population and a decrease in the working-age population in our country, the demographic dividend is gradually disappearing, and labor costs are continuously rising. Therefore, the ad-vantages of automated delivery robots in significantly reducing labor costs and improving delivery efficiency have become extremely important. These factors create favorable conditions for the large-scale application of automated delivery robots. However, the current indus-trial chain of automated delivery robots has not yet formed a complete system, and

S.-H. Sheu (Ed.): IEIM 2024, CCIS 2070, pp. 112–128, 2024.
https://doi.org/10.1007/978-3-031-56373-7_10

its stability and reliability still need improvement. This has led to an unclear attitude of the general public towards automated delivery robots, thereby limiting the further development of this industry.

1.1 Literature Review and Research Hypotheses

The delivery robots studied in this research refer to the robots used in the last-mile delivery service, where the "last mile" refers to the distance from the logistics sorting center to the customer's hands. The satisfaction of users largely depends on the quality of this process and the interaction experience with the delivery robots. Sound is an important aspect of service contact communication [2]. The attractiveness of employees' voices significantly affects customers' expectations and satisfaction with service contact [3]. Similarly, the sound attributes of robots are a crucial aspect in human-robot interaction [4]. Previous studies have shown that robot voice design has a significant impact on the human-machine interaction process. Eyssel found that participants prefer to anthropomorphize robots psychologically when using human-like voices, and this perceived anthropomorphism leads to more positive responses from participants in human-machine interaction [5].Sims et al. pointed out that in the presence of robots with speech capabilities, participants give more commands to synthetic speech robots than to natural speech robots because participants perceive robots with human-like voices to be more capable, thus requiring fewer commands [6]. When robots use human-like voices, it evokes more pleasure and less negative attitude from humans [7]. Furthermore, research has indicated that being able to speak with synthetic or natural voices is sufficient for a robot to be perceived as a competent agent [8]. On one hand, voice design allows users to perceive intelligence, increases trust, and enhances user acceptance [9]. On the other hand, the robot's voice is a major factor influencing consumers' positive emotions, with emotion being the primary mechanism through which the robot's voice affects consumer outcomes [10].

When the robot's emotions vary, people's attitudes and interactions with the robot also change, and the robot's positive emotions can influence the user's emotions [11]. Several studies have explored the design of robot's voice and body language, gaze, and gestures to induce different emotional states in robots. It has been found that the emotional state of the robot does impact human actions, decisions and cooperation [12]. The expression of emotions by robots during cooperation may make team members prefer the robot. So we have made the following assumptions:

H1: Delivery robots with speech have higher user satisfaction than delivery robots without speech.

H2: Delivery robots with high emotional speech have higher user satisfaction than delivery robots with low emotional speech.

Additionally, a "matching hypothesis" [13] has been proposed to explain how people determine the suitability of a robot for performing tasks based on its appearance. Some studies have extended this hypothesis to include the combination of robot sound and appearance. User preferences for robot voice design vary in different service environments and types. Xiao et al. [14] explored the most popular voice types in different application fields and found that the ideal voice type depends on the robot's task content. For a robot used as a shopping receptionist, the most acceptable voice is that of an adult male or a child. For a home companion robot, adult male voices or children's voices are suitable. For robots used in education, adult male and adult female voices are the most acceptable. Previous studies have shown that robots with humanoid features are more likely to elicit social reactions and are therefore more acceptable [15]. Limited research in the field of robotics and artificial intelligence suggests that factors such as personalization can enhance human trust in robots and their sense of responsibility. It has emphasized the decisive role of personification among the various factors that affect the service experience of service robots. Personification characteristics, whether in form or behavior, can promote interaction between people and robots [16]. However, opposite results indicate that, under certain conditions and for various reasons, users may have a negative reaction to the personification of robots. The Uncanny Valley is often used to explain this negative reaction. According to the theory, as the humanoid appearance of a robot becomes more similar to humans, people's response to the humanoid robot changes from acceptance to rejection, and may even lead to negative attitudes such as feeling weird, fear, or a threat to human identity perception [17]. Recent studies have also confirmed this theory. Ferrari et al. [18] conducted research on group differences in social psychology and found that if a robot closely resembles human appearance, blurring the boundaries of identity and weakening human uniqueness, concerns about humans and potential damage to identity may intensify, leading to a negative attitude towards the robot. Therefore, we make the following assumptions:

H3: Under conditions of high emotional speech, delivery robots with a high humanoid appearance have higher user satisfaction.

H4: Under conditions of low emotional speech, delivery robots with a low anthropomorphic appearance have higher user satisfaction.

1.2 Willingness to Pay and Electroencephalogram (EEG) Technology

In this study, user experience is measured by willingness to pay (WTP). WTP refers to the monetary cost that consumers are willing to pay for a certain good or service. Phillips [19] suggests that WTP reflects the degree to which consumers are willing to pay for goods or services that meet their needs. In this way, you can focus on exploring the needs and values of goods, which is also suitable for investigating the attitudes and experiences of users towards the delivery robot that is not yet popular. Event-related potential (ERP) is a neuroscience technology that has high temporal resolution and can be used to evaluate the temporal process of brain activity and potential neural mechanisms in social decision-making [20]. While most previous studies have focused on subjective reporting of user experience and willingness to pay, few have explored the neural mechanism [21]. Subjective reports may be influenced by emotions and societal expectations. Additionally, observer bias may affect participant responses, making it difficult to reflect participants' immediate and exact feelings during human-machine interaction service evaluations, thus introducing bias to the overall results.

While traditional questionnaires or focus groups do not yield accurate results when it comes to understanding consumer attitudes and behaviors because they are influenced by factors such as memory, verbal ability, and social perceptions, neuro-marketing can be derived from signals (related to emotions, attention, memory, etc.) directly from the consumer's brain, which is the most cutting-edge neurological knowledge and technology available. Technology. It can help us understand very accurately what tastes, sounds, colors and smells appeal to our target audience through fMRI, EEG, eye-tracking, skin conductance and other methods. Since delivery robots are not yet relatively popular, mainstream questionnaire methods are difficult to investigate in real-life scenarios and cannot accurately obtain users' attitudes. Therefore, this study adopts the EEG technique, which can be accurate to 1 ms in time for brain activity, which belongs to the most commonly used measurements in neuromarketing nowadays, and can be used to well study the users' emotional and behavioral responses during the decision-making process of their willingness to pay after seeing the appearance of delivery robots and hearing their language.

Electroencephalogram (EEG) technology captures data in real-time and dynamically during the user experience, objectively measuring the user experience based on the brain's cognitive process and personal perception. This makes it convenient for exploring the neural mechanism of the delivery robot's willingness-to-pay process. Previous studies have shown that prefrontal asymmetry in the gamma band and trends in the beta band recorded during product viewing are significantly correlated with subsequent willingness-to-pay responses [22]. Additionally, Jones et al. found that when consumers with higher Maslow's hierarchy of needs made the "purchase" decision, the price-induced P2 amplitude without promotion was larger [23]. The P3 family and LPP are late positive components. P3 is an endogenous component that is generally induced at the task stage and is related to the internal decision-making process. N2, P2, and late positive component (LPP) can be used as indicators of satisfaction, and both frontal and central regions show left-hemisphere advantages for satisfactory web page inter-face processing [24]. Therefore, this study will use ERP technology to investigate the impact of emotional intensity of speech on users' willingness to pay, and examine the neural mechanisms involved in the decision-making process of willingness to pay.

2 Method

2.1 Participants

A total of twenty-two college students (12 females and 10 males) aged from 18 to 25 years old, with an average age of 23.5 years old, participated in the within-subject experiment. The experiment had a 3 (voice type: high-emotion-degree speech, low-emotion-degree speech and warning tone) *2 (appearance type: high-personification and low-personification) design. The data of two participants were discarded due to mistakes made by the experimenter during the research process, resulting in a final sample of 20 participants (10 females and 10 males) with an average age of 23.6 years old. All participants had normal or corrected visual acuity, nor-mal hearing, and no reported history of mental or neurological problems. They were all right-handed. The study was approved by the Ethics Committee of Huaqiao University, and all participants provided informed consent and received compensation after the experiment.

2.2 Experimental Materials

The sound materials used in this paper were selected from recordings of 18 professional actresses. Each actress recorded speeches with high emotional degree and low emotional degree. These recordings were then scored by an expert group based on the level of arousal, and the recording material with the largest difference in arousal level was selected [25]. The recorded sentences consisted of 26 sentences related to express delivery in the service field, with each sentence containing 3–8 words and lasting no more than 2 sec. When selecting non-verbal sound materials, commonly heard message tones and very harsh recordings were excluded, and ultimately, 23 tones within 2 sec in duration were chosen. All sound materials were digitally recorded using Adobe Audition with a constant gain level, a sampling rate of 44100 Hz, a resolution of 16 bits, and unity in decibels.

Since the number of existing prototype delivery robots is limited, the researchers initially selected the robot prototype used in this research from the ABOT database. The researchers selected the prototypes based on their humanoid appearance, using a scale ranging from 0 to 100 for each dimension. When selecting the robots, the researchers prioritized those that were capable of delivering goods, including the developed and used delivery robots like dru and cassie. We selected 15 robot prototypes with a high overall human similarity score (30–100 points) and 15 robots with a low overall human similarity score (0–30 points). Following the method that Sacino described [26], all images were converted to grayscale using the open-source software Krita and pasted onto a white background. The images were then synthesized into an image of 300×400 pixels. An example of the standardized stimuli for the robots used in each experiment is shown in Fig. 1.

(a)   **(b)**

Fig. 1. Example of robot pictures: (a) example of low anthropomorphic robot pictures; (b) example of high anthropomorphic robot pictures

2.3 Experimental Procedures

The experiment utilized a 3 (speech type: high affective language, ground affective speech, and prompts) × 3 (delivery robot anthropomorphism level: high anthropomorphism, low anthropomorphism) within-subjects design. The experiments were run on the E-prime software package, and the entire experiment consisted of 240 trials that were pseudo-randomly assigned to 4 blocks, and the order in which all conditions of the experiment (for a Block) were presented at a time was randomized. Prior to the start of the experiment, participants were informed about the context of the experiment and were asked to imagine a scenario in which they received a delivery service from a delivery robot, and to rate their willingness to pay based on the experience of visual and auditory interactions during the service. During the experiment, subjects sat comfortably in a chair, approximately 75 cm from a computer monitor, and wore TANCHJIM oxygen headphones. Prior to the formal experiment, each subject was required to complete 5 practice sessions to familiarize themselves with the experimental task. The experiment was divided into 4 blocks (each containing 60 trials), with a 2-min break before moving on to the next block. As shown in Fig. 2, the back-ground color was gray during the experiment. At the beginning of each trial, an 800 ms " + " gaze point indicated the start of the experiment. Then, a 2200 ms segment of the robot's voice was presented, followed by a 500 ms gaze point, followed by a 1000 ms picture display of the delivery robot. Finally, a willingness-to-pay scoring screen was presented for the scoring task, which was not time-limited and used a seven-point Likert scale ranging from 1 (very reluctant to pay) to 7 (very willing to pay), with subjects clicking directly on the screen to rate their satisfaction level with the delivery robot's voice and appearance [27].

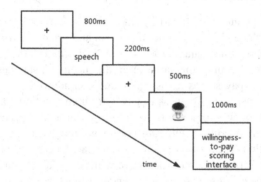

Fig. 2. Experimental structure diagram

2.4   Electroencephalogram (EEG) Recording and Analysis

The willingness to pay was recorded not only through scale scoring, but also through electroencephalography (EEG) recordings. EEG data was collected using the SynAmps 8050 model EEG system produced by Neuroscan and a cap containing 64 Ag/AgCl electrodes. The system operated at a sampling rate of 1000 Hz and followed the standard 10–20 configuration. Participants were comfortably seated and informed about the safety of the experiment. EEG caps were placed on their heads, with electrodes M1 and M2 serving as reference electrodes at the mastoid process behind the left and right ears, respectively. Finally, four electrodes (VEOU, VEOL, HEOL, and HEOR) were placed to remove ocular artifacts from the EEG signals. The impedance of each electrode was maintained below 10 kΩ during the experiment. The EEG data were analyzed offline using Curry8 and MATLAB. Baseline correction and low-pass filtering below 30 Hz were applied to the EEG data, with a time window of 200 ms before and 800 ms after the participant's decision to start. Artifacts exceeding ±75 μV were excluded. The overall event-related potentials (ERPs) data were obtained by stacking and averaging the processed data.

This study primarily focused on the participants' decision-making stage, which was related to willingness to pay. Based on previous research, we analyzed two ERP components: P300 and LPP. The average amplitudes within the time window of 250–400 ms and 400–700 ms, respectively, were examined. For analysis, electrode positions FZ, CZ, and PZ at the midline, as well as F3, F4, C3, C4, P3, and P4 at the bilateral positions, were selected. Repeated-measures analysis of variance was conducted to examine the effects of willingness to pay and electrode position at the midline, and willingness to pay, hemispheric lateralization, and electrode position at both sides. All data were statistically analyzed using SPSS 26.0, with a significance level of 0.05. The p-values for the analysis of variance were corrected using the Greenhouse-Geisser method.

3 Results

3.1 Behavioral Results

The resulting dataset contained 20 participant scores for 240 trails, with six conditions and 40 scores in each condition, and the final mean and standard error were calculated by combining all participant data, 4,800 in total, and utilizing the standard formulas in SPSS 26.0. Figure 3 depicts the variation in the mean and standard error of participants' willingness-to-pay scores across three audio conditions. There are significant differences in participants' willingness to pay, with high emotional speech conditions eliciting greater willingness to pay compared to low emotional speech conditions, and low emotional speech conditions eliciting greater willingness to pay compared to the warning tone condition. The results of a one-way repeated measures analysis of variance revealed significant differences in the degree of willingness to pay under the conditions of high emotional voice, low emotional voice, and warning tone ($F = 104.850$, $p < 0.001$, partial Eta square $= 0.056$). Bonferroni multiple mean comparison results indicated that the degree of willingness to pay for high emotional voice was significantly higher than that for low emotional voice and warning tone ($p < 0.001$), and the degree of willingness to pay for low emotional voice was significantly higher than that for warning tone ($p < 0.001$). Figure 4 displays the changes in willingness to pay under three voice conditions and two appearance conditions. The results showed that the anthropomorphism of the appearance of delivery robots did not significantly affect the impact of speech type on participants' willingness to pay. Regardless of whether the robot had a high or low anthropomorphic appearance, the willingness to pay was highest for high-emotion speech, and in the case of tones, the score for the high anthropomorphic appearance robot was significantly lower than that for the low anthropomorphic appearance robot. The results of repeated measures ANOVA are shown in Table 1, indicating that both speech type and appearance have a significant impact on willingness to pay, with a significant interaction effect between the two.

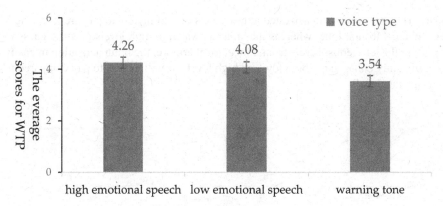

Fig. 3. Influence of voice type on user's willingness to pay

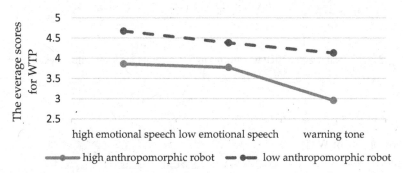

Fig. 4. The influence of speech type and appearance type on user's willingness to pay

Table 1. The results of repetitive measures ANOVA for speech type and appearance type

Type	F	p	Partial η^2
Voice type	116.382	<0.001	0.117
Appearance type	284.883	<0.001	0.245
Voice * appearance	22.149	<0.001	0.025

3.2 Electroencephalogram Results

As shown in Fig. 5, P3 and LPP components were observed in all three levels of willingness to pay when making judgment decisions. This paper analyzed the P3 and LPP components with time windows of 250–400 ms and 400–700 ms. The brain topography maps in Figs. 6 and 7 demonstrate clear differences in brain activation induced by different levels of willingness to pay. According to the questionnaire results in the experiment, 1–3 points are classified as low willingness to pay, 4 points are classified as medium willingness to pay, and 5–7 points are classified as high willingness to pay. There is significant left-right hemi-sphere asymmetry between low and high levels of willingness to pay. The areas primarily activated at low levels of willingness to pay are in the right hemisphere's frontal lobe, whereas the areas activated at high levels of willingness to pay are in the left hemisphere's frontal lobe. Furthermore, the brain activation intensity is significantly enhanced at both low and high levels of willingness to pay compared to inter-mediate levels.

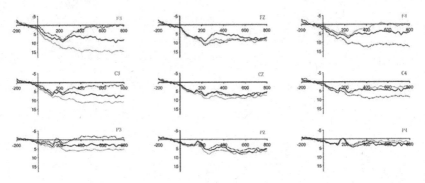

Fig. 5. The total average ERP waveforms under different levels of willingness to pay

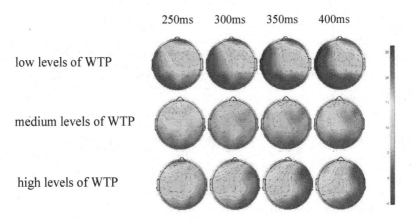

Fig. 6. The brain topography maps of P300 under different levels of WTP

3.3 LPP

The LPP components observed in the time window of 400–700 ms were analyzed. A three-factor repeated measures analysis of variance was conducted to examine the effects of willingness to pay (high, medium, and low), hemisphere (left side, right side), and pole placement (frontal, central, and apical regions) on the two sides of the brain (F3/F4, C3/C4, and P3/P4). The results are presented in Table 2.

The table reveals the following findings: The main effect of willingness to pay was not significant (F = 0.514, p = 0.602). The main effect of cerebral regions was significant (F

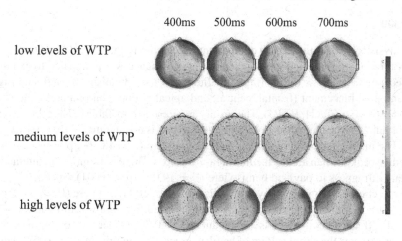

Fig. 7. The brain topography maps of LPP under different levels of WTP

Table 2. The results of the three-factor repeated analysis of variance of LPP components

type	F	p	Partial η^2
Voice type	116.382	<0.001	0.117
Degree of willingness to pay	0.514	0.602	0.026
Brain region	48.360	<0.001	0.718
hemisphere	0.706	0.498	0.036
Degree of willingness to pay × hemisphere	25.391	<0.001	0.572
WTP × brain region × hemisphere	15.068	<0.001	0.442

$= 48.360, p < 0.001$), with the average peaks being highest in the frontal region, followed by the central region, and then the apical region. The interaction between willingness to pay and hemispheres was significant ($F = 25.391, p < 0.001$). In the left hemisphere, the simple effect of willingness to pay was significant ($F = 8.349, p = 0.003$), and in the right hemisphere, the simple effect of willingness to pay was also significant ($F = 5.214, p = 0.016$). Under the condition of low willingness to pay, the average peak value in the left hemisphere was smaller than that in the right hemisphere ($p = 0.003$). Under the condition of high willingness to pay, the left hemisphere showed larger values than the right hemisphere ($p < 0.001$). Under the condition of medium willingness to pay, there was no significant difference between the left and right hemispheres. The interaction among willingness to pay, cerebral region, and hemisphere was significant ($F = 15.068, p < 0.01$). The simple effect of willingness to pay was significant in both the left and right hemispheres of the frontal region, central region, and apical region ($p < 0.01$).

3.4 P300

The analysis focused on the P300 components occurring in the 250–400 ms time window. A three-factor repeated measures analysis of variance was conducted to examine the effects of willingness to pay (high, medium, and low), hemisphere (left side, right side), and pole placement (frontal, central, and apical regions) on bilateral large brain electrodes (F3/F4, C3/C4, P3/P4). The results are presented in Table 3. The table reveals several findings: The main effect of willingness to pay is not significant ($F = 1.130$, $p = 0.328$). The main effect of cerebral regions is significant ($F = 48.360$, $p < 0.001$), with the average peak median area > median area > top area. There is a significant interaction between willingness to pay and hemisphere ($F = 19.069$, $p < 0.001$). Specifically, the simple effect of willingness to pay is significant in the left hemisphere ($F = 6.242$, $p = 0.009$), but not in the right hemisphere. Under the condition of low willingness to pay, the simple effect of hemisphere is significant ($p = 0.012$). Under the condition of high willingness to pay, the simple effect of hemisphere is also significant ($p = 0.007$). However, under the condition of medium willingness to pay, the simple effect of hemisphere is not significant.

The interaction effects of willingness to pay degree, cerebral regions, and hemispheres are significant ($F = 14.528$, $p < 0.001$). The simple effect of willingness to pay is significant under the conditions of left and right hemispheres in the frontal region, central region, and left hemisphere in the apical region ($p < 0.05$). In summary, the analysis of variance revealed significant effects of cerebral regions, hemisphere, and their interactions on the P300 components. The specific patterns of these effects are detailed in Table 3.

Table 3. Three-factor repeated measures analysis of variance of p300 component

type	F	p	Partial η^2
Degree of willingness to pay	1.130	0.328	0.056
Brain region	77.314	$p < 0.001$	0.803
Degree of willingness to pay × hemisphere	19.069	$p < 0.001$	0.501
Willingness to pay × brain region × hemisphere	14.528	$p < 0.001$	0.433

4 Discuss

The results of this study indicate that the satisfaction level of users with voice-enabled delivery robots is higher than that of users without voice-enabled delivery robots, which supports Hypothesis 1 and is consistent with previous research findings. Previous studies have shown that participants issue more commands to synthetic voice robots in the presence of voice functionality because they perceive voice-enabled robots to be more capable, thus confirming the positive impact of robots' ability to speak on users [28]. Most of the previous researches are focused on social robots, and through our experiments we can also prove that the design of voice function is not only very important for social robots, but also essential for delivery robots, and the incorporation of voice design will largely increase the users' willingness to pay for delivery robots, which is an important factor for the promotion and adoption of de-livery robots. Therefore, it can be concluded that the inclusion of speech design will enhance users' willingness to pay for delivery robots. Additionally, the results of this study also suggest that users' satisfaction level is higher with delivery robots that have high emotional voice compared to those with low emotional voice, supporting Hypothesis 2. Consistent with previous research findings, experimental results demonstrate that robots' use of emotional strategies in voice has high persuasiveness, and positive robot emotions can indeed influence users' emotions [29]. On one hand, individuals are able to discern a robot's emotions through its voice or facial expressions; Dubal et al. [30] revealed that people can accurately distinguish a robot's emotional state with a success rate of 94%, and when individuals encounter a robot exhibiting happiness, the amplitude of the early P100 component is higher. On the other hand, different emotions displayed by robots elicit varying responses from individuals. Reuten [31] discovered that when robots exhibit negative emotions, people's pupils dilate the most. Moreover, in behavioral experiments involving human-robot interactions, individuals demonstrate a tendency to trust and engage in friendlier communication [32] with robots able to express positive emotions. Previous studies have examined the impact of robot speech on users' experience and perception of user behavior through behavioral experiments, with results suggesting that the quality of voice, intonation, and modes of emotional expression all influence user satisfaction and subsequently affect their willingness to use [33]. It is of particular note that, in contrast to previous research, this study emphasizes that the emotion of robot voices is not only reflected in language strategies, but also in the emotional tone of the voice. This is a key focus of the study, as even slight variations in voice emotion can impact user experience while maintaining consistency in language content and style. The results of the experiment further confirm the findings of Appel et al. [34], demonstrating that robots with a higher degree of emotional expressiveness are more popular than those with neutral emotion.

In terms of the interaction design of voice and appearance, users consistently exhibit higher willingness to pay for delivery robots with low anthropomorphic appearance compared to those with high anthropomorphic appearance, supporting Hypothesis 4, but not Hypothesis 3. Although the level of anthropomorphism in robots' perception affects the positivity of emotional experience, thereby influencing users' acceptance of robots, excessive anthropomorphism can lead to negative attitudes towards robots [35]. The findings of relevant studies suggest that for robots with different degrees of human likeness, people tend to experience feelings of joy and satisfaction towards highly anthropomorphic robots [36]. Both appearance and speech can enhance the anthropomorphism of robots, and as the overall level of anthropomorphism increases, users' trust and willingness to pay also increase [37]. Some studies have indicated that it is easier for humans to establish emotional connections with humanoid robots [38] compared to non-humanoid robots [39]. From the experimental results, it can be observed that under conditions of high anthropomorphic appearance, the difference in willingness to pay between high and low emotional voice is indeed larger compared to other situations. In addition, the experimental results show that the impact of appearance anthropomorphism on user experience exceeds that of emotional voice level. Under the condition of low anthropomorphic appearance, users exhibit higher willingness to pay for delivery robots with high emotional speech compared to those with low emotional speech, once again highlighting the significant influence of emotional voice level on user experience. Previous studies have indicated that the sound of robots is a primary factor influencing consumers' positive emotions. Moreover, when the level of anthropomorphism in robot voices is higher, they are rated as more pleasant and less frightening, receiving the highest acceptance scores across various practical application scenarios. In the emotion recognition system developed by Hong, robots with expressive abilities are more frequently engaged in interaction with human users. The positive emotions conveyed by the robots consequently influence users' emotions. Notably, the voice of the robot serves as the primary means of displaying its emotions, followed by body language and eye color. These findings align with the results of the present study. While previous studies have focused on the impact of appearance or voice alone, this study not only considers the impact of voice affectivity on user experience, but also combines different anthropomorphic degrees of appearance design at the same time, to better explore the moderating effect of voice affectivity on the user's visual experience, as well as the impact of matching degree of voice affectivity and appearance anthropomorphism on the user's willingness to pay. In this study, we combined the voice and appearance of delivery robots and found both the difference in the priority of their impact on user experience and the moderating effect of voice emotionality on appearance anthropomorphism.

In scientific and industrial research, the neurological approach has been demonstrated as an excellent predictor of population behavior. Neural scoring has shown to be superior to reaction time in predicting population behavior, while reaction time, in turn, outperforms questionnaire surveys. In recent years, there has been an increasing number of studies utilizing electroencephalography (EEG) to investigate user experience and willingness to pay in human-computer interaction. These studies include testing consumer willingness to pay for different products, examining the independent neuropsychological mechanisms underlying consumer choices and dynamically measuring emotions during the user experience process. All of these studies indicate that the methods of neuroscience can alleviate some biases present in subjective questionnaires and allow for WTP (willingness to pay) studies with smaller sample sizes than traditional methods. It has been found that even in relatively small samples, brain responses can predict not only individual choices but also market effects. Neuroscience methods can be applied not only to test the impact of product attributes on WTP but also to evaluate the WTP effects of specific messages in marketing communication. Thus, they can provide a better understanding of consumer behavior and be used to assess consumer preferences and choices. Additionally, in relevant research, scholars have recorded the human-machine interaction process between users and humanoid robots through physiological methods. This is done to observe the fundamental mechanisms of social cognition triggered by humanoid robots. Guo Fu et al. [40] also assessed user preferences for robot voices through EEG experiments. Therefore, the method of electroencephalography (EEG) proves highly suitable for the present study, as well as future related research endeavors.

The findings from the Event-Related Potential (ERP) investigation conducted in this study indicate that that making willingness-to-pay decisions triggers the P300 component and LPP component. It is generally believed that the P300 component is related to the allocation of attentional cognitive resources, and the greater the stimulus intensity, the higher the level of arousal of the subjects, leading to the investment of more attentional cognitive resources [41]. In the research results, there is no significant relationship between willingness to pay and average amplitude of ERP, but there is a significant interaction effect of brain region and hemisphere in the P300 component. In the left hemisphere, the simple effect of willingness to pay is significant. The LPP component shows obvious asymmetry between the left and right hemispheres in the case of low and high levels of willingness to pay. Previous research has shown that willingness to pay is positively related to the amplitude of late positive potential, and higher levels of willingness to pay will elicit larger amplitudes [42]. In addition, LPP and P300 have been shown to be related to decision-making processes in related work and can be used as an indicator of user satisfaction. Larger LPP responses are closely associated with strong emotional stimuli, reflecting higher emotional arousal. LPP reflects a broader allocation of attention due to the affective and motivational relevance of the pictures, which leads to higher amplitudes corresponding to more arousing stimuli [43]. LPP amplitude is enhanced when participants are immersed in emotional rather than neutral scenes. These findings are consistent with the results of the present study, where participants' LPP activation was significantly lower during the process of making a medium willingness-to-pay decision than during the process of making a high willingness-to-pay decision. The emotional reactions in the purchase decision-making process are also supported by another study, which found that product prices activate brain regions responsible for emotion processing, and product prices and rating cues serve as emotional stimuli, triggering different emotional arousals [44]. This study, on the other hand, adds that the emotional stimulus that triggers users' affective and cognitive processes in payment decision-making can be speech with different degrees of affectivity, and the experimental results have well demonstrated that language with different degrees of affectivity has significantly different levels of emotional arousal for the process of willingness to pay.

5 Conclusion

This study utilized ERP technology to investigate the impact of speech emotion on users' willingness to pay for automatic delivery robots from the perspective of speech interaction. The results showed that, regardless of whether the delivery robot had a low anthropomorphic appearance or a high anthropomorphic appearance, speech with high emotional intensity was more effective in increasing users' willingness to pay compared to speech with low emotional intensity. Additionally, delivery robots with speech capabilities were found to better meet users' needs compared to those without speech capabilities. The study also found a significant correlation between the emotional intensity of speech and users' willingness to pay, which was supported by changes in the P3 and LPP components measured through EEG signals during the WTP decision-making stage. These findings provide a new way for businesses to predict and influence consumers' purchasing decisions. They also offer new insights and methods for the design and marketing of future delivery robots.

This study has some limitations. Firstly, our study only considered the emotionality attribute of speech, while other attributes of speech may have an impact on user experience and willingness to pay. Future research should expand the research scope in terms of voice attributes, such as naturalness of voice and gender. Second, our study only focused on the degree of willingness to pay for automated delivery robots among the college student population in China. Considering the differences in the degree of knowledge and awareness of technology and automated delivery robots among people of different age groups, as well as the cultural differences among different countries and regions, our detailed experimental design and process can provide references for studying people in other regions and other age groups. Third, the delivery and payment scenarios simulated in the lab are different from the real scenarios, and the performance of the participants may be different as a result. Future research can further explore the attitudes of the users in hotels, shopping malls, restaurants, and other environments where automated delivery robots are more widely used, using questionnaires and interviews, and using more realistic VR scenarios in combination with EEG instruments.

In future research, we will further explore other stages of willingness to pay by employing different research paradigms and ERP components. We will examine the impact of robot speech and appearance on cognitive processes and subsequent decision-making processes in the stages of presenting delivery robots with different types of speech and appearances. With the widespread use of delivery robots, it is possible to explore user behavior and experiences during face-to-face interactions with delivery robots in real-life scenarios. This will contribute to a deeper understanding of the process by which users' willingness to pay for delivery robots is formed, and provide more guidance for improving user experience and market competitiveness.

References

1. Pani, A., Mishra, S., Golias, M.: Evaluating public acceptance of autonomous delivery robots during COVID-19. Transportation Research Part D, pp. 3–45 (2020)
2. Burgers, A., de Ruyter, K., Keen, C., Streukens, S.: Customer expectation dimensions of voice-to-voice service encounters: a scale-development study. Int. J. Serv. Ind. Manag. 11(2), 142–161 (2000)
3. Bartsch, S.: What sounds beautiful is good?" How employee vocal attractiveness affects customer's evaluation of the voice-to-voice service encounter. Current Research Questions in Service Marketing, pp. 45–68 (2008)
4. Moon, B.J., Choi, J., Kwak, S.S.: Pretending to be okay in a sad voice: social robot's usage of verbal and nonverbal cue combination and its effect on human empathy and behavior inducement. In: 2021 IEEE/RSJ International Conference on Intelligent Robots and Systems (IROS), pp. 854–861 (2021)
5. Eyssel, F., De Ruiter, L., Kuchenbrandt, D., Bobinger, S., Hegel, F.: 'If you sound like me, you must be more human': on the interplay of robot and user features on human-robot acceptance and anthropomorphism. In: Proceedings of the 7th ACM/IEEE International Conference on Human-Robot Interaction (HRI 2012), pp. 125–126, March 2012

6. Tussyadiah, I.P., Park, S.: When guests trust hosts for their words: host description and trust in sharing economy. Tour. Manage. **67**, 261–272 (2018)
7. Li, M., Guo, F., Wang, X., Chen, J., Ham, J.: Effects of robot gaze and voice human-likeness on users' subjective perception, visual attention, and cerebral activity in voice conversations. Comput. Hum. Behav. **141**, 107645 (2023)
8. Sims, V.K., Chin, M.G., Lum, H.C., Upham-Ellis, L., Ballion, T., Lagattuta, N.C.: Robots' auditory cues are subject to anthropomorphism. In: Proceedings of the Human Factors and Ergonomics Society Annual Meeting, vol. 53, pp. 1418–1421 (2009)
9. Tay, B., Jung, Y., Park, T.: When stereotypes meet robots: the double-edge sword of robot gender and personality in human–robot interaction. Comput. Hum. Behav. **38**, 75–84 (2014)
10. Lu, L., Zhang, P., Zhang, T.C.: Leveraging "human-likeness" of robotic service at restaurant. Int. J. Hospitality Manage. (2021)
11. Hong, A., et al.: A multimodal emotional human-robot interaction architecture for social robots engaged in bidirectional communication. IEEE Trans. Cybern. **51**(12), 5954–5968 (2021)
12. Takahashi, Y., Kayukawa, Y., Terada, K., Inoue, H.: Emotional expressions of real humanoid robots and their influence on human decision-making in a finite iterated prisoner's dilemma game. Int. J. Soc. Robot. **13**, 1777–1786 (2021)
13. Torre, I., Latupeirissa, A.B., McGinn, C.: How context shapes the appropriateness of a robot's voice. In: Proceedings of the 2020 29th IEEE International Conference on Robot and Human Interactive Communication (RO-MAN) (2020)
14. Dou, X., et al.: Effects of different types of social robot voices on affective evaluations in different application fields. Int. J. Soc. Robot. **13**(4), 615–628 (2021)
15. Eyssel, F., Kuchenbrandt, D.: Manipulating anthropomorphic inferences about NAO: the role of situational and dispositional aspects of effectance motivation. In: Proceedings of the RO-MAN 2011, pp. 467–472, July 2011
16. Duffy, B.R.: Anthropomorphism and the social robot. Rob. Auton. Syst. **42**(3–4), 177–190 (2003)
17. Mori, M.: The uncanny valley. Energy **7**(4), 33–35 (1970)
18. Ferrari, F., Paladino, M.P., Jetten, J.: Blurring human–machine distinctions: anthropomorphic appearance in social robots as a threat to human distinctiveness. Int. J. Soc. Robot. **8**(2), 287–302 (2016)
19. Phillips, K.A., Homan, R.K., Luft, H.S., Hiatt, P.H.: Willingness to pay for poison control centers. J. Health Econ. **19**, 343–357 (1997)
20. Li, M., Li, J., Li, H., Zhang, G., Fan, W., Zhong, Y.: Interpersonal distance modulates the influence of social observation on prosocial behaviour: An event-related potential (ERP) study. Int. J. Psychophysiol. **176**, 108–116 (2022)
21. Jun, J., et al.: Research on user experience evaluation of mobile games based on EEG technology. J. Inf. Syst. **2**, 38–52 (2018)
22. Liu, Y., Li, F., Tang, L.H.: Detection of humanoid robot design preferences using EEG and Eye tracker.In: Proceedings of the 2019 International Conference on Cyberworlds (CW), pp. 219–224 (2019)
23. Jones, W.J., Childers, T.L., Jiang, Y.: The shopping brain: math anxiety modulates brain responses to buying decisions. Biol. Psychol. **89**(1), 201–213 (2012)
24. Guo, F., et al.: Research on evoked potential of satisfaction evaluation of web interface. Ind. Eng. Manage. **21**(03), 126–131 (2016)
25. Chen, X., Yang, X., Yang, Y.: The neurophysiological mechanism of implicit processing of phonological emotional changes. J. Psychol. **45**(04), 416–426 (2013)
26. Sacino, A., et al.: Human- or object-like? Cognitive anthropomorphism of humanoid robots. PLoS ONE **17**(7), 0270787 (2022)

27. Peykarjou, S., et al.: Audio-visual priming in 7-month-old infants: an ERP study. Infant Behav. Dev. **58**, 1–9 (2020)
28. Saunderson, S., Nejat, G.: Investigating strategies for robot persuasion in social human-robot interaction. IEEE Trans. Cybern. **52**(1), 641–653 (2022)
29. Traeger, M.L., Strohkorb Sebo, S., Jung, M., Scassellati, B., Christakis, N.A.: Vulnerable robots positively shape human conversational dynamics in a human–robot team. Proc. Natl. Acad. Sci. **117**(12), 6370–6375 (2020)
30. Ramsøy, T.Z., Skov, M., Christensen, M.K., Stahlhut, C.: Frontal brain asymmetry and willingness to pay. Front. Neurosci. **12**, 138 (2018)
31. Zhang, G., Li, M., Li, J., Tan, M., Li, H., Zhong, Y.: Green product types modulate green consumption in the gain and loss framings: an event-related potential study. Int. J. Environ. Res. Public Health **19**(17), 10746 (2022)
32. Torre, I., Goslin, J., White, L.: If your device could smile: people trust happy-sounding artificial agents more. Comput. Hum. Behav. **105**, 106215 (2020)
33. Verhulst, N., Vermeir, I., Slabbinck, H., Lariviere, B., Mauri, M., Russo, V.: A neurophysiological exploration of the dynamic nature of emotions during the customer experience. J. Retail. Consum. Serv. **57**, 102217 (2020)
34. Appel, M., Izydorczyk, D., Weber, S., Mara, M., Lischetzke, T.: The uncanny of mind in a machine: humanoid robots as tools, agents, and experiencers. Comput. Hum. Behav. **102**, 274–286 (2020)
35. Zhang, Y., Wang, Y.: Research on the mechanism of the impact of service robot anthropomorphism on consumer use intention - The regulatory role of social class. Foreign Econ. Manage. **44**(03), 3–18 (2022)
36. Seo, S.: When Female (Male) Robot Is Talking To Me: Effect of service robots' gender and anthropomorphism on customer satisfaction. Int. J. Hosp. Manag. **102**, 103166 (2022)
37. Yoganathan, V., Osburg, V.S., Kunz, W.H., Toporowski, W.: Check-in at the Robo-desk: Effects of automated social presence on social cognition and service implications. Tour. Manage. **85**, 104309 (2021)
38. Riek, L.D., Rabinowitch, T.-C., Chakrabarti, B., Robinson, P.: How anthropomorphism affects empathy toward robots. In: Proceedings of the 4th ACM/IEEE international conference on Human robot interaction. ACM, La Jolla, California, USA, pp. 245–246 (2009)
39. Bartneck, C., Bleeker, T., Bun, J., Fens, P., Riet, L.: The influence of robot anthropomorphism on the feelings of embarrassment when interacting with robots. Paladyn **1**, 109–115 (2010)
40. Li, M., Guo, F., Chen, J., Duffy, V.G.: Evaluating users' auditory affective preference for humanoid robot voices through neural dynamics. Int. J. Hum. Comput. Interact., 1–19 (2022)
41. Nieuwenhuis, S., Aston-Jones, G., Cohen, J.D.: Decision making, the p3, and the locus coeruleus-norepinephrine system. Psychol. Bull. **131**, 510–532 (2005)
42. Hongjuan, S., Yushi, J., Wei, L.: Knight's dynamic price sentiment under uncertainty: a study based on Evoked potential. Manage. Sci. **34**(01), 113–129 (2021)
43. Chatterjee, A., Mazumder, S., Das, K.: Reversing food preference through multisensory exposure. PLoS ONE **18**(7), 1–26 (2023)
44. Sun, L., Zhao, Y., Ling, B.: The joint influence of online rating and product price on purchase decision: an EEG study. Psychol. Res. Behav. Manag. **13**, 291–301 (2020)

Optimization of Dyeing Process of Cotton Fibers in a Peruvian Textile Company

Fabiana Crosato-Oyarce⬤, Michelle Sauter-Olivé⬤,
and Marcos Fernando Ruiz-Ruiz$^{(\boxtimes)}$⬤

Faculty of Engineering, Industrial Engineering Career, Universidad de Lima, Lima, Perú
{20190548,20191902}@aloe.ulima.edu.pe, mruiz@ulima.edu.pe

Abstract. The textile industry is one of the most polluting industries in the world, due to the significant environmental impact it generates. Therefore, the adoption of measures to optimize its processes is necessary to reduce production times, as well as water and energy consumption. In this regard, this study conducted an analysis of cotton fiber production at ITESSA, a Peruvian textile company. The main objective was to optimize the dyeing process of cotton fibers using the DMAIC continuous improvement methodology. Through its implementation, it was identified that the dye used caused multiple production delays, leading to a proposal for a change to the Grupo CHT brand dye, which allows dyeing with fewer baths and at a lower temperature without compromising color quality and intensity. With its application, the dyeing time was reduced by 10 h, water consumption by 7500 L, and energy usage by 38 kW-h, accompanied by economic savings for the company. The methodology presented here can be replicated in other contexts and similar companies.

Keywords: cotton · textile sector · productivity · dyeing process · DMAIC methodology

1 Introduction

The textile industry is one of the most polluting industries in the world. In recent years, textile production has increased [1] due to population growth, improved living standards [2], and the trend of 'buying and discarding' [1]. As a result, the amount of textile waste has also expanded, leading to greater pollution in recent past [3].

This global issue is reflected in the consumption and pollution of water resources, as the textile industry consumes about 93 million cubic meters of water annually and releases half a million tons of microfibers into the ocean [4]. Additionally, 700,000 tons of dyes are used to dye fabrics and garments, of which 84,000 tons are discharged into rivers, seas, and oceans without prior purification processes [5]. Thus, the textile industry is responsible for 20% of the worldwide wastewater generated [4]. This environmental pressure has led to the search and adoption of measures to optimize dyeing processes, aiming to reduce production times, water consumption, and energy usage [6].

The main objective of the study was to optimize the dyeing line, specifically in the production of yarns made from cotton fibers, to improve productivity in a Peruvian

© The Author(s), under exclusive license to Springer Nature Switzerland AG 2024
S.-H. Sheu (Ed.): IEIM 2024, CCIS 2070, pp. 129–141, 2024.
https://doi.org/10.1007/978-3-031-56373-7_11

textile company. This was achieved by reducing production times and costs, leading to an economic improvement for the company upon implementation.

1.1 Cotton Fiber in the Textile Industry

Cotton is the most useful and valuable natural fiber in the textile industry [6]. It stands out for being soft with unique properties of durability, strength, and absorbency [6], making it suitable for a variety of fabrics and garments. Its high demand in the textile industry makes cotton a raw material of great economic importance [7]. In Peru, both Pima and Tangüis cotton are cultivated, known as some of the most prized varieties in the world [8].

The textile sector utilizes cotton as a raw material because garments made from this are known for their comfort [9]. This sector is divided into the textile industry and the garment industry. The former encompasses processes from cotton treatment for yarn production to fabric manufacturing, while latter focuses on garment assembly [10].

Although cotton is considered an easily dyeable fiber [9], Campo et al. [11] propose that it is important to focus optimization efforts on processes such as dyeing, as it is often a bottleneck activity, and its improvement can make a significant difference in productivity. Thus, optimization is defined by Campo et al. [11] as the action aimed at improving the quality or reducing costs in the production process of cotton fibers. Similarly, Ocampo [12] refers to it as the analysis of a process to improve inefficient activities and elements that cause delays.

Finally, the productivity of the cotton yarn line is defined as the ratio between the production of yarn and the resources used to obtain it [13]. Peñaloza et al. [14] also emphasize the relationship between productivity, process performance, and efficient resource utilization to maximize the quantity of cotton yarn produced. Productivity increases as production grows and inputs are maintained [13].

1.2 Diagnosis of the Problem

Based on the observed issues, a study was conducted on Industrias Textiles de Sudamérica (ITESSA), a company in the textile sector that works with a wide variety of fibers. According to their financial manager, the main problem identified in the study is the high production cost caused by the lengthy dyeing process, with cotton fiber being particularly notable, taking an average time of 18 to 23 h [15].

To corroborate the information provided by the manager, a block diagram (see Fig. 1) was created, considering the capacity of vat 3, which can handle 30 kg. Additionally, the machine's capacity and the established batch size were used to determine the time in minutes required for each activity.

In the block diagram it was identified that the dyeing process is the bottleneck. This activity starts with the preparation of the cotton fibers, followed by neutralization and dyeing of the fiber. Then, the cotton is neutralized again and finishes with soaping and softening.

In the dyeing process curve of ITESSA, several idle times were observed, where it is necessary to wait for the temperature to reach the 98 °C required for dyeing. Additionally,

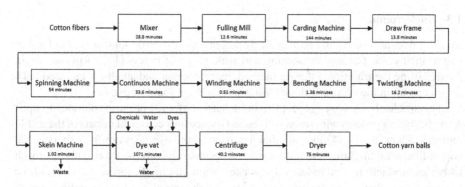

Fig. 1. Block diagram of the cotton yarn production process

different neutralizing and auxiliary agents are required for color fixation, which need to be added at different temperatures. On the other hand, the water in the vats is replaced multiple times, which increases the cost of water resources and leads to more idle times (see Fig. 2).

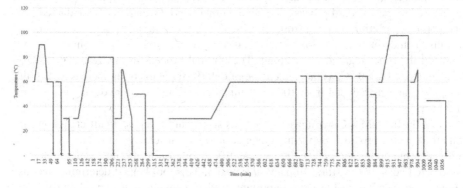

Fig. 2. Current dyeing process curve of ITESSA

Additionally, a detailed investigation was conducted on the dyeing processes in other factories in the Peruvian textile sector, revealing a significant technical gap with the process used by ITESSA. Bohorquez's research [16] indicates that the dyeing of cotton fibers takes between 5 to 13 h. Similarly, Aragón [6] and Ocampo [12] indicate that the duration of the dyeing process in other companies varies between 8.05 and 8.7 h. Therefore, it was concluded that on average, the dyeing of cotton fibers takes between 8 and 9 h, while for ITESSA, it took between 17 and 20 h. Additionally, Aragón [6] mentions that 10 rinses are necessary for the cotton dyeing process, whereas ITESSA requires 18 machine discharges. This indicates that ITESSA consumes 80% more water than other companies in the sector.

1.3 Background

Reactive dyes are commonly used in the dyeing process of textile companies and fashion due to their color fastness in washing and wide range of colors [17]. However, to fix the dye using the traditional dyeing method, inorganic salt and alkali are required [17], which end up in the effluents along with a significant amount of dye [18].

The articles by Acharya et al. [19] and Khatri et al. [17] propose solutions to optimize the dyeing process with reactive dyes and reduce the ecological impact of the textile industry. Khatri et al. [17] indicate that a portion of the dye does not fully fix onto the cotton fibers, highlighting the importance of optimizing color fixation. This approach increases productivity and reduces dye waste. Proposed modifications include machine modifications, changes to the dyeing process, and the use of biochemics in fiber preparation [17]. Acharya et al. [19] proposed using ethanol to reduce the amounts of salt and water used in cotton dyeing. The study resulted in increased color fixation and intensity, optimizing the process while reducing the consumption of salt, water, and dyes [19]. Regarding the soaping activity, research has been conducted to evaluate the quality of the subprocess at different temperatures. In the article by Perugachi [7], the aim was to remove hydrolyzed dye, and three tests were performed in the soaping process at different temperatures: 100 °C, 90 °C, and 80 °C. After the study, it was observed that the optimal temperature for this activity is 80 °C, as it improved color intensity by 2.18% compared to the standard process [7]. On the other hand, Ocampo [12] conducted four temperature variations ranging from 60 °C to 90 °C. These variations showed promising results, as they increased colorfastness values in cotton, improved color intensity, and reduced the time required for soaping. The optimal process in this research was the test conducted at 60 °C, as the percentage of colorfastness in cotton increased, and the soaping time was reduced from 100 to 35 min, decreasing the entire dyeing process by 1 h [12].

Finally, the DMAIC optimization tool has stood out for its use in different contexts and the benefits provided by its implementation. In the article by Ibarra and Berrazueta [20], this methodology was applied to identify a solution to the high defect rate and associated costs in a textile company in Ecuador. This tool helped determine the customers' perspective on the most important product characteristics and identify flaws in the weaving and printing process. As a result, they were able to identify improvement opportunities and changed the washing process, reducing the percentage of quality defects by 1.86% [20]. Additionally, Rojas [21] used the DMAIC methodology to improve productivity in the washing process of a textile company. The application of this tool, along with necessary audits, resulted in a 15% increase in the production of the company analyzed in the research [21]. Based on these two articles, it can be concluded that the DMAIC tool is a successful instrument for continuous improvement that allows for the identification of problems and efficient optimization of production processes.

2 Methodology

The research was defined as a case study with an evaluation of productivity indicators before (pre-test) and after (post-test) the implementation of an improvement proposal for the cotton fiber dyeing process at ITESSA. A pre-experimental design was used since the study worked with a pre-defined fiber and in a specific area of a Peruvian company chosen based on accessibility criteria.

The selected methodology followed the principles of the DMAIC philosophy, given the benefits found in similar cases to address the identified issues [16]. The study began by analyzing and measuring the dyeing activities to identify problems, design solutions, and optimize the process [22]. It concluded with measuring the optimized process and comparing results obtained in both the pre-test and post-test phases. The methodology employed a mixed approach, as information from the company was collected both quantitatively and qualitatively to generate the improvement proposal.

Data collection was conducted through interviews with individuals currently working in the factory and direct observation on the plant floor. Also, we obtained support from the production manager of the dyeing line, who provided detailed information about the process and demonstrated its functioning. Additionally, during the walkthrough, necessary indicators for comparative analysis were identified and collected. Finally, the Microsoft Excel® program was used to record, process, and analyze the data collected between the pre-test and post-test phases.

As mentioned earlier, the DMAIC methodology was used to implement the proposal. This tool allowed for the improvement of quality and productivity in the dyeing area by identifying potential enhancements and prioritizing the most relevant ones to achieve an optimal process time (see Fig. 3).

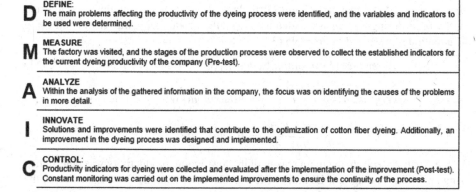

Fig. 3. DMAIC Process Diagram

In the Define component, the dyeing area process was analyzed in detail, and possible limitations were evaluated, such as high dyeing temperatures, high water and energy consumption, and idle times. Afterwards, all variables and indicators to be used in the measurement and data collection were defined (see Table 1).

For the Measure component, the indicators listed were calculated. For it to be done, the dyeing vat 3 of ITESSA was taken as a reference, which allows dyeing 30 kg of fibers with a water capacity of 750 L. The activity times were obtained with the assistance of the dyeing supervisor, who measured the hours it took to perform each activity using a stopwatch. The costs were determined during the interview with the production manager,

Table 1. Indicators for pre-test and post-test to evaluate the productivity of the textile fiber dyeing line.

Variables	Indicators	Unit of Measurement	Productivity
Time	Preparation time	Minutes (min)	Kilograms/Minutes (kg/min)
	Dyeing time		
	Soaping and softening time		
Cost	Total cost of inputs in preparation	Dollars (USD)	Kilograms/Dollars (kg/USD)
	Total cost of inputs in dyeing		
	Total cost of inputs in soaping and softening		
Water consumption	Liters of water used	Liters (L)	Kilograms/Liters (kg/L)
Energy consumption	Kilowatts of energy used	Kilowatts (kW)	Kilograms/Kilowatts (kg/kW)

who provided information about the process inputs. Using the established technical specifications of the machinery and Formulas (1) and (2), the amount of water and energy used during dyeing was calculated. Finally, productivity was calculated using the equation represented in Formula (3).

$$Water\ used\ (L) = Bathtub\ capacity\ (L) \times Number\ of\ baths \tag{1}$$

$$Energy\ used\ (kW) = Energy\ consumption\ (kW-h) \times Time\ (h) \tag{2}$$

$$\frac{Producto}{Insumo} = Productivity \tag{3}$$

For the Analyze section, the main reason of the low dyeing productivity was determined using an Ishikawa diagram. Ultimately, it was identified that the main issue lied in the type of dye used and the conditions it requires for application.

In the Innovate component, an improvement proposal was designed where new possibilities of dyeing with various dyes were identified. After selecting the appropriate one, the optimization was conducted in the factory using a sample of 30 kg of cotton fiber to validate the improvement, and the optimized dyeing curve was determined.

Finally, in the Control segment, the same indicators used in the Measure stage were collected, and the post-test indicators were determined. Next, the productivity increase was calculated using Formula (4), and constant monitoring of the improvements was conducted to ensure the continuity of the process.

$$\frac{p_2 - p_1}{p_1} \times 100\% \tag{4}$$

3   Results

The results were obtained by analyzing a sample of cotton fiber in dyeing vat 3 of ITESSA. This machine has a capacity of 30 kg of fiber, 750 L of water, and requires 3.5 kW-h of energy. The dye used in the measurement was black (see Table 2).

$$Water\ used\ (L) = 750L \times 18Baths = 13500L$$

$$Energy\ used\ (KW) = 3.5\,kW\text{-}h \times 17.85\ h = 62.475\,kW$$

Table 2. Pre-test Results

Variables	Indicators	Values	Productivity
Time	Preparation time	351 min	$\frac{30kg}{1071min} = 0.028kg/min$
	Dyeing time	539min	
	Soaping and softening time	181min	
Cost	Total cost of inputs in preparation	15.28 USD	$\frac{30kg}{82.15USD} = 0.37kg/USD$
	Total cost of inputs in dyeing	56.08 USD	
	Total cost of inputs in soaping and softening	10.79 USD	
Water consumption	Liters of water used	13500 L	$\frac{30kg}{13500L} = 0.0022kg/L$
Energy consumption	Kilowatts of energy used	62.5 kW	$\frac{30kg}{62.5kW} = 0.48kg/kW$

Afterwards, the analysis of the Ishikawa diagram identified in detail the main causes of the dyeing problem, which were high energy and water consumption, as well as high costs (see Fig. 4). It was observed that the type of dye used is the issue that generates all the root causes due to its specific usage requirements.

After identifying the problems an improvement proposal was designed and validated. This proposal involved changing the dyes used and restructuring the dyeing process to align with the requirements of the new dye. Changing to a dye that works at lower temperatures is ideal for process improvement because it reduces the idle time caused by waiting for temperature increases and rinsing. It also directly reduces energy costs since the machine needs fewer hours of operation to achieve the same result, and there is no need for high energy consumption to heat the 750 L of water. Various dyes and brands were investigated, such as Kisco, Coltex, and Grupo CHT.

Kisco was discarded as their dying process required temperatures ranging from 60 °C to 130 °C [23]. These elevated temperatures would maintain the current issues of

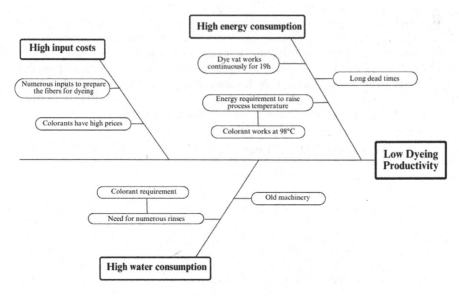

Fig. 4. Ishikawa Diagram

ITESSA regarding the numerous waiting times and energy requirements. Additionally, Coltex was not considered for improvement because the company does not provide technical specifications for the dye or the working temperature [24]. Finally, a new type of dye used in the market, Bezaktiv Go [25], was identified. This dye from the Grupo CHT does not require a significant temperature increase and allows for dyeing and rinsing the fibers at 40 °C. These dyes were used in the improvement design due to their advantageous characteristics and usage requirements. Bezaktiv Go also "requires minimal consumption of water, energy, […] and dye, […] without compromising color intensity and process safety" [25]. It was decided to obtain a quote and apply the Bezaktiv Go dye, as well as the recommended inputs, to optimize the dyeing process and improve ITESSA's dyeing model using 30 kg of cotton fiber with black dye (see Fig. 5).

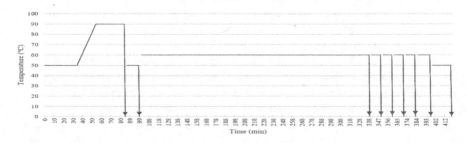

Fig. 5. Optimized dyeing curve

After implementing the improvement, the same dyeing vat and indicators were used to determine the post-test results (see Table 3).

$$Water\ used\,(L) = 750L \times 8Baths = 6000L$$

$$Energy\ used\ (KW) = 3.5\ \text{kW - h} \times h = 24.5\ \text{kW}$$

Table 3. Post-test Results

Variables	Indicators	Values	Productivity
Time	Preparation time	100 min	
	Dyeing time	245 min	$\frac{30\,kg}{420\,min} = 0.071\,kg/min$
	Soaping and softening time	75 min	
Cost	Total cost of inputs in preparation	14.5 USD	$\frac{30\,kg}{71.54\,USD} = 0.42\,kg/USD$
	Total cost of inputs in dyeing	47.53 USD	
	Total cost of inputs in soaping and softening	9.51 USD	
Water consumption	Liters of water used	6000 L	$\frac{30\,kg}{6000\,L} = 0.005\,kg/L$
Energy consumption	Kilowatts of energy used	24.5 kW	$\frac{30\,kg}{24.5\,kW} = 1.22\,kg/kW$

Finally, a comparative table of the calculated productivities in the pre-test and post-test was prepared to identify the productivity increase achieved with the proposed optimization. It also helps determine the extent to which the technical gap is closed with the applied improvement (see Table 4).

Table 4. Comparison of pre-test and post-test productivity indicator

Variables	Pre-test	Post-test	Increase in productivity
Time	0.028 kg/min	0.071 kg/min	$\frac{0.071\,kg/min - 0.028\,kg/min}{0.028\,kg/min} \times 100\% =$ 153.57%
Cost	0.37 kg/USD	0.42 kg/min	$\frac{0.42\,kg/USD - 0.37\,kg/USD}{0.37\,kg/USD} \times 100\% =$ 13.51%

(continued)

Table 4. (*continued*)

Variables	Pre-test	Post-test	Increase in productivity
Water consumption	0.0022 kg/L	0.005kg/L	$\frac{0.005\,\text{kg/L} - 0.0022\,\text{kg/L}}{0.0022\,\text{kg/L}} \times 100\% =$ 127.27%
Energy consumption	0.48 kg/kW	1.22 kg/kW	$\frac{1.22\,\text{kg/kW} - 0.48\,\text{kg/kW}}{0.48\,\text{kg/kW}} \times 100\% =$ 154.17%

4 Discussion

The main factors contributing to the technical gap in the process are time, temperature, and the number of dye baths. It is crucial to consider all three aspects in order to optimize the dyeing process and narrow the technical gap with the rest of the industry.

Regarding the reduction in time, the dyeing process was reduced from 17.85 h to 7 h, increasing productivity from 0.028 kg/min to 0.071 kg/min, representing an increase of 153.57%. Although the proposed time in the study by Bohorquez [16] was not achieved, it closely resembled the dyeing time proposed in the investigations by Aragón [6] and Ocampo [12], where the duration was 8.05 and 8.7 h, respectively. This increase in productivity benefits ITESSA and helps close the technical gap. The optimization will also lead to a reduction in labor costs by $21.42 per dyeing load, leveraging the savings of 10.85 h in the process. Additionally, the company will be able to dye an additional 0.05 kg of cotton fiber per dollar invested, resulting in a savings of $9.65 per production load. Overall, considering that each load is 30 kg, the company would save $31.07 in variable costs.

The optimization of dyeing temperature is extremely important as it affects the salt dissolution, color intensity, and colorfastness of the cotton [12]. In the proposed improvement, the dyeing temperature was reduced from 98 °C to 60 °C, as it still provides the necessary temperature for salt dissolution and allows for a shorter dyeing process by eliminating the need to heat water. This aligns with the suggestion by Ocampo [12] that the optimal dyeing temperature is 60 °C. Furthermore, it improves the color intensity by 2.18% compared to the standard process, increases the colorfastness of cotton, and reduces the time required for soaping from 100 to 35 min, saving 1 h of the process. On the other hand, Perugachi [7] proposes that the optimal temperature is 80 °C as it enhances the colorfastness of cotton, improves color intensity, and reduces the time required for soaping. Although both projects showed positive results, the decision was made to use the 60 °C temperature proposed by Ocampo, as the chemistry of Bezaktiv Go dyes allows for a more intense and deep shade at that temperature.

Using a dye that operates at a lower temperature is beneficial as it requires less energy and shortens the process by eliminating the need to heat water. Moreover, the change in dye reduces the number of dye baths, thereby reducing water usage. The initial dyeing process of ITESSA had 18 dye baths, requiring 454 L of water per kg of cotton fiber, while the proposed improvement has 8 water drains and a consumption of 200 L per kg of fiber. This increases productivity by 127.27% and surpasses the proposal by Aragón

[6], which indicates the need for 10 rinses in cotton dyeing. As a result, ITESSA goes from consuming 80% more water to 20% less than other companies in the sector.

Finally, the study had some limitations. Firstly, there was no consideration for machinery replacement, as it would require a larger investment that the company was not willing to make in the short term. Secondly, the study focused solely on cotton fibers, while ITESSA works with a much wider range of fiber types such as alpaca, recycled fibers, and blends utilizing different fibers. Thirdly, although the factory has two sizes of dyeing vats (30 kg and 200 kg), the sample was conducted with a small batch of 30 kg. This limitation is noteworthy as variables and dyeing quality can vary depending on the batch size. Lastly, the sample used black colorants, so there was no need to obtain an exact shade, and color differences would be imperceptible to the human eye.

5 Conclusions

After conducting the research, it is observed that the DMAIC methodology plays a significant role in the optimization process. With this tool, it was identified that the colorant initially used by ITESSA was a common factor contributing to the low productivity in the dyeing process, prompting all strategies to focus on obtaining a better-quality colorant. Additionally, the implementation of the colorant change optimized the dyeing of cotton fibers and reduced the processing time from 1071 to 420 min. It is important to continue analyzing this activity and seeking further improvements as it remains the bottleneck activity.

The use of colorants that operate at lower temperatures also proved beneficial for the dyeing process of cotton fibers as it eliminates the need for elevated temperature increases, resulting in lower energy consumption and reduced process times. This benefited the company as it managed to decrease its electricity consumption by 54% and the process time by 53.5%.

The implementation of CHT Group colorants and the recommended inputs in the pre-dyeing stage led to a reduction in water consumption. By making these changes, ITESSA eliminated 8 unnecessary baths in its process, reducing its water consumption by 27%. Although the input costs were higher, there is a significant benefit in implementing these inputs as they increase process productivity and reduce company expenses.

It is important to consider the purchase of a small machine with a propeller to facilitate salt dissolution and reduce the sub-process time. This is necessary because salt dissolution is currently done manually, and it becomes challenging to achieve proper salt dissolution with the decreased temperature in the dyeing process.

This work serves as a starting point in the research line of innovation, technology, and products. It is suggested to continue the study for the application of other fibers such as nylon, alpaca, and recycled fibers. Furthermore, the transfer of the presented methodology to evaluate different colorants and their effects on dyeing productivity is proposed. Future studies could also evaluate the efficiency of machinery in this type of industry.

References

1. Riba J.-R., Cantero, R., Canals, T., Puig, R.: Circular economy of post-consumer textile waste: classification through infrared spectroscopy. J. Cleaner Prod. **272**(123011), 9 (2020). https://doi.org/10.1016/j.jclepro.2020.123011
2. Wang, Y.: Fiber and textile waste utilization. Waste Biomass. Valor. **1**(1), 135–143 (2020). https://doi.org/10.1007/s12649-009-9005-y
3. Dahlbo, H., Aalto, K., Eskelinen, H.: Increasing textile circulation—Consequences and requirements. Sustain. Prod. Consumption **9**(23525509), 44–57, 2017. https://doi.org/10.1016/j.spc.2016.06.005
4. Naciones Unidas: Datos y cifras—Naciones Unidas. The United Nations (2019). https://www.un.org/es/actnow/facts-and-figures
5. Artifon, W., Cesca, K., de Andrade, C.J., Ulson de Souza, A.A., de Oliveira, D.: Dyestuffs from textile industry wastewaters: trends and gaps in the use of bioflocculants. Process Biochem. **111**, 181–190, 2021. https://doi.org/10.1016/j.procbio.2021.10.030
6. Aragón Vallenas, J.C.: OPTIMIZACIÓN Y REDUCCIÓN DE COSTOS DEL PROCESO DE TEÑIDO DE TEJIDOS DE POLIÉSTER/ALGODÓN SIN ALTERAR LA SOLIDEZ DEL LAVADO (2012). Google Académico. https://web.archive.org/web/20180503193104id_/http://cybertesis.uni.edu.pe/bitstream/uni/3343/1/aragon_bj.pdf
7. Perugachi Vásquez, C.J.: OPTIMIZACIÓN DEL PROCESO DE TINTURA DE ALGODÓN 100% CON LOS COLORANTES TINA BEZATHREN A NIVEL LABORATORIO EN LA EMPRESA TEXTIL "QUIMICOLOURS S.A.", 10 November 2017. Google Académico. http://repositorio.utn.edu.ec/handle/123456789/7410
8. Industrias Textiles de Sud América SAC. (n.d.): Compañía. ITESSA. https://www.itessa.com.pe/es/compania
9. Xia, L., et al.: Environmentally friendly dyeing of cotton in an ethanol–water mixture with excellent exhaustion. Green Chem. **20**(19), 4473–4483 (2018). https://doi.org/10.1039/c8gc0181
10. Court, E., Pérez, V., Rodríguez, C., Ingar, B.: Reporte Financiero Burkenroad Perú – Sector Textil del Perú, 27 September 2010. Google Académico. http://artesaniatextil.com/wp-content/uploads/2017/04/BRLA-Peruvian-Textile-Industry-201003.pdf
11. Campo, E.A., Cano, J.A., Gómez-Montoya, R.A.: Optimization of aggregate production costs in textile companies. Revista chile de ingeniería **28**(3) (2020). https://www.scielo.cl/scielo.php?pid=S0718-33052020000300461&script=sci_arttext
12. Ocampo Dávila, S.S.: Optimización del proceso de teñido reactivo de tejidos de algodón sin afectar la apariencia y la solidez al lavado, en el área de tintorería de una empresa textil localizada en Lima-Perú (2019). Google Académico. http://cybertesis.unmsm.edu.pe/handle/20.500.12672/10758
13. Prokopenko, J.: LA GESTIÓN DE LA PRODUCTIVIDAD (1) (1989). Google Académico. https://d1wqtxts1xzle7.cloudfront.net/38639804/Libro-Productividad-Prokopenko-libre.pdf?1441160835=&response-content-disposition=inline%3B+filename%3DGestion_de_la_productividad.pdf&Expires=1681794412&Signature=Xq2XEpRydPkegt8ZAVFS5j6W-78bYgX1EIlKJ2PtU6SlY
14. Peñaloza Arredondo, F., Hernández Razo, J.C., Llamas Pérez, G.A.: Diseño de distribución eficiente de planta para el aumento de la productividad en la empresa Grupo T&M. Universidad de Guanajuato **3**(1), 537–540 (2017). https://www.jovenesenlaciencia.ugto.mx/index.php/jovenesenlaciencia/article/view/925/pdf1
15. Crosato, A.: Entrevista con el gerente general de ITESSA (2022)

16. Bohorquez Avila, G.A.: ANÁLISIS DE LA MEJORA DEL PROCESO DE TEÑIDO MEDIANTE LA APLICACIÓN DE LA METODOLOGÍA LEAN SIX SIGMA EN UNA EMPRESA TEXTIL DE LA CIUDAD DE AREQUIPA, 2021 (2022). Google Académico. https://repositorio.ucsm.edu.pe/bitstream/handle/20.500.12920/12182/44.0806. II.pdf?sequence=1&isAllowed=y

17. Khatri, A., Peerzada, M.H., Mohsin, M., White, M.: A review on developments in dyeing cotton fabrics with reactive dyes for reducing effluent pollution. J. Cleaner Prod. **87**(1), 50–57 (2015). https://doi.org/10.1016/j.jclepro.2014.09.017

18. Shu, D., Fang, K., Liu, X., Cai, Y., Zhang, X., Zhang, J.: Cleaner coloration of cotton fabric with reactive dyes using a pad-batch-steam dyeing process. J. Cleaner Prod. **196**, 935–942 (2018). https://doi.org/10.1016/j.jclepro.2018.06.080

19. Acharya, S., Abidi, N., Rajbhandari, R., Meulewaeter, F.: Chemical cationization of cotton fabric for improved dye uptake. Cellulose **21**(6), 4693–4706 (2014). https://doi.org/10.1007/s10570-014-0457-2

20. Ibarra Albuja, C.D., Berrazueta Lanas, G.S.: Aplicación metodología DMAIC en empresa textil con enfoque en reducción de costos (2019). Google Académico. https://repositorio.usfq.edu.ec/handle/23000/8174

21. Rojas Apolinario, G.: Implementación de la metodología DMAIC para mejorar la productividad del proceso de lavado textil en la Empresa Industria Textil del Pacifico S.A. 2016 (2016). Google Académico. https://repositorio.ucv.edu.pe/handle/20.500.12692/3435

22. Ruiz-Ruiz, M.F., Diaz-Garay, B.H., Noriega-Aranibar, M.T.: Gestión e investigación en ingeniería: revisión sistemática de literatura para Iberoamérica. Revista Venezolana de Gerencia **27**(98), 597–618 (2022). https://doi.org/10.52080/rvgluz.27.98.14

23. Kisco. (n.d.): Colorantes - Kisco. Kisco. https://www.kisco.com.pe/colorantes/

24. Coltex. (n.d.): Inicio - Coltex Peru. Coltex Peru SAC—Químicos y Colorantes. http://coltexperu.com.pe/

25. Grupo CHT. (n.d.): BEAZAKTIV GO—Colorantes sostenibles y eficientes - Grupo CHT - Químicas especiales. CHT. https://solutions.cht.com/cht/web.nsf/id/pa_bezaktiv_go_fokus_es.html

Productivity Enhancement by Layout Redesign and Application of Lean Principles in the Apparel Industry

Leslie Fernandez-Diaz[ID], Natalia Vera-Rojas[ID], and Juan Carlos Quiroz-Flores[✉][ID]

Faculty of Engineering, Industrial Engineering Career, University of Lima, Lima, Peru
{20173345,20174026}@aloe.ulima.edu.pe, jcquiroz@ulima.edu.pe

Abstract. The growth of the textile-apparel industry has been significantly harmed by the pandemic experienced in 2019. However, this was compounded by previous problems such as the low quality of products, the large amount of waste, and unproductive times, among others. These have had a significant impact on the low productivity of this sector, which hinders its ability to compete with large Asian competitors. An essential part is the correct management of processes and plant design. The proposed proposal uses tools such as TQM, 5S, Poka-yoke, SLP, and process standardization to improve the current productivity indicator of 0.04 und/PEN after the implementation of the pilot plans and the simulation carried out with Arena, an increase in the secondary and primary indicators was obtained, which reached 0.11 und/PEN.

Keywords: Applied computing · Transportation · Supply Chain Management

1 Introduction

The textile industry is an essential sector in the world economy, which is currently going through challenging circumstances given the competitive environment in search of competitive prices, excellent quality, and innovative solutions that ensure the reduction of operating costs to compete with foreign manufacturers such as Asia [1]. In this sense, low productivity and delivery delays are essential factors in Peruvian SMEs' need for more development [2]. For this country, the textile and apparel industry contributed to 2.3% of national employment, 0.8% of the national Gross Domestic Product (GDP), and 6.4% of the manufacturing GDP in 2019, despite the drop in the last decade of 0.9% in the participation of the national economy and 4.2% in the manufacturing GDP [3]. Likewise, within the sub-activities of the sector, the apparel sector registered a 35.9% drop in its production level for 2020 due to the restriction of operations, a lower internal and external demand, and the large entry of imported garment volumes [4]. That is supported by the BCRP (Central Reserve Bank of Peru), which indicates that the utilization rate of the sector's installed capacity has fallen to 42.6% by the end of 2020, resulting in lower production activity [5]. That shows that Peruvian SMEs have some operational restrictions that limit production and affect the proper growth of the company to face large foreign competitors.

© The Author(s), under exclusive license to Springer Nature Switzerland AG 2024
S.-H. Sheu (Ed.): IEIM 2024, CCIS 2070, pp. 142–154, 2024.
https://doi.org/10.1007/978-3-031-56373-7_12

The problem identified, according to the literature, may be due to high production times, high number of reprocesses, and high amount of waste, among others, which leads to low productivity of the companies [6]. This problem was also perceived in research in other countries and regions. For example, in a textile company in India, only 44 garments/day/human resources could be manufactured due to a deficient line balance and a large amount of waste generated, which resulted in increased activity times that did not create value and did not allow the product to be delivered on time [7]. Another research in Pakistan identified that the productivity of a textile company could have been higher due to high changeover times and waste in the production process, which resulted in excessive waste of time that affected efficiency and productivity [8]. Without going far, in Peru, in a study carried out in a textile company, low productivity was identified as affected by a large amount of waste and cycle time, which affected the efficiency of material consumption and, therefore, productivity [9]. As mentioned above, it is shown that textile companies have inefficiencies in their production processes, which leads to high operating costs, and, therefore, in productivity, so new solutions to this problem are ideal.

In this context, Peruvian SMEs must be efficient by eliminating waste and increasing production. To this end, a case study reflects the sector's problem of low productivity due to different types of waste. Those identified were the high reprocesses and wastage rate due to defects and bottlenecks, which generated 20.41% of total sales economic losses. To solve the problem, Lean Manufacturing was implemented under the SMED, Visual Control, and Standardization tools. That was created using success stories discovered in the literature study, and it meets the requirement to address the issue and advance science. This research aims to investigate the Lean Manufacturing work models in companies of the textile sector, given the scarce information presented, especially in the South American continent. It is intended to demonstrate that through this tool, significant changes can be achieved in Peruvian textile companies, especially in micro and small enterprises (MSEs) that monopolize a large part of the market and that would benefit by improving their productivity through low investment, thus reducing operating overheads by integrating all the company's resources and identifying those activities that do not add value.

The scientific article has eight parts: Summary, Introduction, Diagnosis, State of the Art, Proposed Model, Validation, Discussion, and Conclusion.

2   State of the Art

2.1 .Production Models to Increase Productivity in the Textile Sector

Productivity plays a strategic role in the ultimate success of an industry [10]. According to studies found, the productivity factor is influenced under the terms of changeover time control utilizing standardized methods [8] to the production index, which developed favorable results by the use of VSM, SMED, and visual control, which allows the identification of bottlenecks in some operations that affect the optimal development of the production process [11], the reduction of cycle times, the efficiency of the use of resources through the 5 s and Line Balancing tools [7], the use of simulation tools for process improvement through the use of JIT, Kanban and process standardization, the percentage of defective parts [2] and the regulation of work shifts through the implementation of SMED, Poka Yoke and standardized work tools [12, p. 8].

2.2 Lean Manufacturing in the Textile Sector

Lean tools allow the establishment of stability and standardization focused on production to improve production performance to incorporate safe and efficient methods to meet quality standards [12, p. 4]. It is evidenced in studies where productivity is improved, continuous improvement is promoted, and waste is minimized [13]. The reduction in production times and defects was evidenced [14]. Reducing waste and process optimization is another factor that positively impacts using Lean tools such as VSM, 5S, TPM, or SMED [6]. In this regard, studies were presented where the positive impact of Lean tools is evidenced as in the case of SMED, process standardization, and Poka Yoke, whose result is evidenced in time management and error reduction by 80%; on the other hand, the efficiency factor was obtained in 87.5%, with an improvement of up to 7% [12, p. 8]. Order, cleanliness, optimal management of workspaces, and increased productivity are other terms considered as part of the continuous improvement of lean tools [15]. Furthermore, the main results of implementing lean tools are improved production line efficiency, inventory reduction, production process improvement, reduction of defective products, and equipment performance [6].

2.3 Total Quality Management in the Textile Sector

Quality management is another of the techniques belonging to the group of Lean tools that significantly impact the performance of the textile company by optimizing the quality of products and services [16]. Some studies expose the relevance of the performance and skills of the workers when elaborating on the products, whose purpose is to become familiar with the new innovative techniques that increase the quality of the products [17]. One of the most used techniques in this management tool is the DMAIC technique, which helped reduce the variation within the processes and quality costs [16]. On the other hand, studies show a significant reduction of defects by almost 80% and an increase in quality by 57.96% [18]. With the Kaizen method, defective products were reduced by 10.19%, which increased product quality and productivity [19]. Also, 20% of the textile industry's waste becomes performance improvement at the organizational level due to the quality culture [20].

2.4 5S Methodology in the Textile Sector

5S methodology originates from the first letter of five Japanese terms: Seiri, Seiton, Seiso, Seiketsu, and Shitsuke. Its application involves sorting items that are not frequently used; arranging tools used daily for easy access; cleaning the workspace; standardizing by developing procedures to ensure compliance with the program and maintaining [15]. This tool positively impacts small and medium-sized companies by reducing manufacturing costs and increasing productivity [21]. Case studies show that applying 5S, the PDCA cycle, and 5W2H saved 4 h per week per operator, which meant a gain of 10% of available time per week [22]. In a case developed in Peru, they applied the 5S and the SLP, which resulted in the reduction of cycle time from 33.64 min per garment to 25.32 min, a decrease of waste from 84% to 16%, reduction of unnecessary movements from 56% to 44% and an increase in income of 7% [23]. Finally, [24], in their research, increased production and reduced the number of stops to 5% and, as a result, obtained elimination of lost time, saving 5,502 min per month. This created 506,184 additional units, an increase of approximately 10.55%.

2.5 Poka Yoke in the Textile Sector

The Poka-Yoke system uses sensors or other instruments mounted on equipment or machinery to track errors that operators overlook. These systems run on two elements: 100% inspection and instant feedback [25]. As a result of mental and physical flaws, human mistakes in production and management processes must be eliminated or reduced. The three main components are eliminating independent mistakes (resistance issues), preventing the sources of errors, and using a reasonably low-cost control system to assess product conformance [26]. Through the use of a control system that keeps the fabric placed in optimal conditions and a change in the closing machine so that the cut minimizes the potential error that can be generated, studies conducted by Poke Yoke have shown the value of the tool that combines quality control methodologies and seeks to cut errors in laying and cutting in a textile company by 80%. The pilot test increased the process' overall efficacy of making pants by 7%, bringing it to 87.5% [12, p. 8].

2.6 Systematic Layout Planning (SLP) in the Textile Industry

Plant layout is a strategic plan that allows companies to organize the elements according to the workflow, allowing tasks to be executed more quickly during production [27]. With this scenario, SLP (Systematic Layout Planning) was developed to develop a procedure to identify, examine and visualize the available areas, machines, and equipment while keeping the work areas more precise [28, p. 2]. This tool solves plant layout problems based on quantitative criteria that seek to increase productivity and reduce costs [29]. The benefits are reflected in the improvement and efficiency of material flow, reduction of cycle times, reduction of unnecessary movements, and clearance of occupied areas [23, p.3]. A study in India found that the material flow path was 1,440 m per bag, and with the implementation of the tool in question, it was reduced to 970 m [28, pp.13–14]. In the same case, in a Peruvian textile SME, the plant travel time was reduced by 13.09% and the effort by 11.12% [23, p. 9]. In another context, the cycle time of 33.64 min per backpack produced was reduced to 25.32 min, and demand fulfillment went from 37% to 86% [23, p. 7].

3   Contribution

3.1   Model Justification

With the analysis of the textile sector, it became evident that low productivity is every day in Peruvian micro, small and medium-sized companies, which has caused their low competitiveness to face large textile producers such as Central America and Asia, being negatively affected in terms of sales, income, quality, and competitiveness. In this sense, we searched for tools, techniques, and methodologies to eliminate the present problem or even improve it. Thus, we found the use of Lean Manufacturing tools, among which are: 5 s, Poka Yoke, standardized work and TQM, and plant layout, as the most used and recommended tools to solve the identified problem. Given the large number of tools involved in Lean Manufacturing, the exploration of the benefits and synergies present among the other tools included remains to be investigated.

3.2   Proposed Model

The use of Lean tools, such as 5s, standardized work, Poka Yoke, and TQM, as well as the use of the Layout redesign methodology, have a positive impact on productivity growth, but, at present, there are a limited number of investigations based on the improvement of small and medium-sized clothing companies through the use of engineering tools and given the structural conditions of small Peruvian companies; an innovative model was designed, which in comparison with other studies conducted, seeks to involve together the use of TQM, SLP, Poka-yoke, process standardization and 5 s methodology tools in a small Peruvian company whose sector is little used by the scientific community, as a means of support for the growth of SMEs in the textile sector for the proper implementation and management of production processes shows the model proposed to the company, with three components incorporating the previously mentioned tools (see Fig. 1).

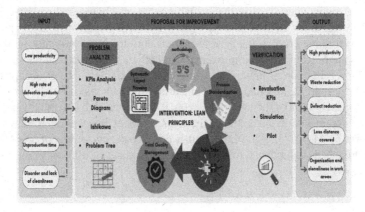

Fig.1. Proposed Model

3.3 Model Components

Component 1: Problem analysis. Information from the firm was gathered for the issue analysis, allowing the formulation of the indicators to be analyzed. The Pareto diagram was used to calculate the effect of the leading causes after developing a Problem Tree to identify the root causes. A case study was created to suggest the instruments for the improvement model.

Component 2: Intervention. This second component is based on applying the five tools in the garment workshop. The first is Systematic Layout Planning (SLP), which will facilitate operators' mobility and improve the current flow of materials in the plant. The aim is to reduce distances traveled and facilitate interchange between areas. The second is the Total Quality Management tool, which aims to reduce the rate of defective T-shirts to optimize the use of material and economic resources. The third tool is the 5 s methodology for reordering and optimizing the storage of materials and products, in which an initial and final internal audit will be carried out. The fourth tool is the Standardization of processes to establish standardized methods to regulate production times and the adequate use of available fabric to reduce waste and shrinkage. The last tool is the Poka Yoke, to reduce in advance the errors present in the cutting area, and avoid reprocesses, defects, and final defective products, which affect the final productivity.

Component 3: Control. In this phase, the values obtained from the indicators used to solve the problems encountered are reevaluated. Furthermore, verify if the objective of improving the company's productivity is met to ensure the model is appropriately executed. After the calculation of the indicators with the new values obtained, a comparison will be made with the values obtained at the beginning to examine the new changes obtained with the use of the tools mentioned above; in this way, the impact originated with the execution of the new model will be verified. In order to evaluate the impact of Lean tools and Layout redesign, the following indicators were proposed (Table 1).

Table 1. Indicators

Indicators	Formula	Use
Productivity	$\frac{Quantity of finished product}{MO+MP+MQ+extras}$	Analyzes the efficiency of using available resources to manufacture a batch of products [19]
Defective product rate	$\frac{Units of defective products}{Total units produced} x100$	Measures the proportion of defective units in a production batch [19]
Waste rate	$\frac{Quantity of wasted materials}{Total quantity of available materials} x100$	Measures the proportion of materials wasted when manufacturing the batch of products [9]

(*continued*)

Table 1. (*continued*)

Indicators	Formula	Use
Efficiency	$\frac{Actual production}{Expected production} x100$	measures resource use efficiency concerning the expected production [19]
Cycle Time	$\frac{The\ sum\ of\ observed\ times}{Number\ of\ cycles\ observed}$	Evaluates the time dedicated to the production of a product [30]
Distance traveled	$\sum Transfer distance$	Measures the distance between workstations to develop productive activities within the plant [23]

4 Validation

4.1 Initial Diagnosis

The present research model aims to study a production workshop of a Peruvian textile SME whose main problem is the technical gap of low productivity of 0.09 und/PEN below the benchmark. For this, it focused on studying the design and production of a printed T-shirt. This problem is caused mainly by the excess of waste and scrap (58.88%) due to the deficient work method and the absence of inspection techniques for the materials; the incorrect disposition of the work areas (28.60%) due to the unnecessary routes made by the operators and the disorder and lack of cleanliness in the work areas, and high unproductive times (12.52%) due to the poor training of the operators and unproductive times during the production process. The economic impact of this problem amounts to S/. 51,785 per year, equivalent to 21.76% of the company's total costs. A problem tree was developed as a tool for identifying root and sub causes (see Fig. 2).

4.2 Design of the Validation and Comparison with the Initial Diagnosis

Three steps served as the foundation for the model validation design. In the first stage, the root causes of the Peruvian company's low productivity were determined by creating a problem tree. It was discovered that 58.88% of the problem is caused by an abundance of waste and scraps in the cutting areas, 28.6% by an uneven distribution of the work areas, and 12.52% by the high rate of unproductive times throughout the manufacturing process. The lean tools were defined by defining their application throughout the next phase. We began by developing the SLP pilot plan, where we designed two redistribution models of the areas where the second option was selected to follow the production sequence. Secondly, the 5 s methodology was applied through an initial audit to then organize and order the warehouses with the 5 s phases and thus execute the final audit of the 5 s and thus show through a graph the variation before and after the application of the 5 s (see Fig. 3).

Then, the guidelines to carry out the TQM were raised, which has the use of registration and control of defective through the intervention of inspection points in cutting,

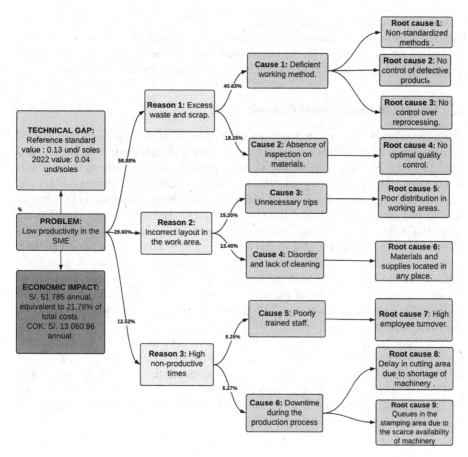

Fig. 2. Problem Tree

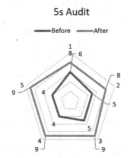

Fig. 3. Audition before and after implementation

sewing, and stamping. The other tool applied is the Poka Yoke, where guides and manuals were established for cutting machines and overlockers to increase efficiency in the production process, as well as a mechanical measuring instrument that allows precise measurements without a margin of error. Moreover, finally, the incorporation of standardized work through sequential guidelines incorporated with images of how to develop the cutting activities; therefore, the last phase focuses on reevaluating the indicators to measure and analyze the evolution of these with the implemented improvement.

4.3 Simulation Improvement Proposal

For the simulation, we chose to model the entire process of making the printed T-shirt, from the arrival of the fabrics purchased by the company already made with the color and type of fabric to be used to the packaging of the printed polo shirt. Among the activities that occupy a large part of the productive time were the cutting and sewing activities. It was also observed, through the timing of the activities, that some operators took longer than others because they followed different techniques to perform the task, as well as the need to have the support of those workers with more time working in the workshop to be able to sew the garments. Faced with this context, we resorted to the use of standardized work and Layout redesign, thus reducing the production time dedicated to those tasks that did not add value to the production process and establishing the correct design of the areas to prevail the order of the activities and eliminate unnecessary routes. For this purpose, Arena simulation software version 16.2 was required, whose simulation is shown (see Fig. 4).

The Input Analyzer program was used to evaluate the data from the 30 observations to get the proper distribution fit for each activity's model. 30 replicates were considered for the computation of the data sample, with a confidence level of 90% and an error of 10%. An ideal size of 87 replicates with a confidence range of [27;54] was found using the current mean width value of 1.72. (see Fig. 4) displays the simulation's findings and the carried-out pilot plan.

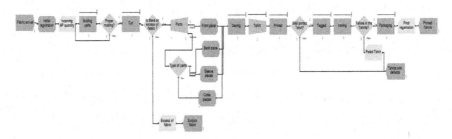

Fig. 4. Simulation Model in Arena

For the development of the analysis of the results, it was required to record the values of the current model in comparison with the values obtained with the improvement formulated according to the tools incorporated for the development of the improvement (Table 2).

Table 2. Comparative table of indicators

Indicator	Initial Value	Objective Value	Final Value
Productivity	0.04 und/PEN	>0.13 und/PEN	0.11 und /PEN
Defective product rate	18.61%	15%	14.86%
Waste rate	28.26%	21.26%	20, 44%
Efficiency	32,10 min	reduce by 5.8%	22,84 min
Cycle Time	84.15%	95%	94.94%
Distance traveled	111.81 m	97.87 m	90.50 m

5 Discussion

The purpose of this improvement model is focused on increasing productivity [2, 19, 32]; and, therefore, reduce the excess of waste and scrap and reduce unproductive times that do not add value to the production process through the use of Lean tools and Layout redesign that help other small companies in the textile sector that present the same problems as those identified in this one, such as the reduction of defective products from 28.26% to 20.44%, which is within the optimal range established by the sector, through the use of TQM with the new techniques of registration and control of defective products [2, 9, 11]. The increase in efficiency of 10.79%, reaching 94.94% but not reaching the expected 95%, through the development of a new layout design for the distribution of work areas and standardized work [12, 19]. The regularization of unproductive cycle times from 32.10 min to 22,84 min, which represents 28.85%, using Poka-yoke, 5 s [6, 12] and standardization of work [24], and finally, the distances traveled decreased from 111.81 m to 90.5 m and the effort from 149037, 60 kg-m/year to 120 969 kg-m/year by implementing the SLP and 5S [23].

6 Conclusion

In conclusion, implementing the proposed model made up of the five tools (TQM, 5s, Poka Yoke, Standardized Work, and SLP) allowed an increase in productivity of 0.8 und/PEN compared to the current value. That occurred thanks to the implementation of standardized work, Poka Yoke, and TQM, which allowed a reduction in the percentage of defective products (14.86%), the percentage of wastage (20.44%), cycle time (28.85%) and efficiency (94.94%). Likewise, by applying the 5S in all the areas involved, order and cleanliness improved, reaching 8.60 points in the final audit. Also, the SLP reduced the distance traveled to 90.50 m, and effort was minimized to 120,969 kg-m/year. Concerning economic impact, reprocessing costs, costs for defective products, extra labor, and costs for PM loss were analyzed. By applying lean principles and SLP, saving a total of S/will be possible 17,394.56, which means an improvement of 28.45% in the economic aspect. On the other hand, for future research, a correct cutting process would avoid possible defects in the subsequent activities and better use of the raw material. Finally, clusters, which are groups of companies belonging to the same type of industry that join to buy and sell products and services among themselves, taking advantage of the proximity of suppliers and buyers to reduce costs, have become more popular.

References

1. Robertsone, G., Mezinska, I., Lapina, I.: Barriers for Lean implementation in the textile industry. Int. J. Lean Six Sigma **13**(3), 648–670 (2022)
2. Canales-Jeri, L., Rondinel-Oviedo, V., Flores-Perez, A., Collao-Diaz, M.: Lean model applying JIT, Kanban, and standardized work to increase the productivity and management in a textile SME. In: Proceedings of the 3rd International Conference on Industrial Engineering and Industrial Management, pp.79–84. IEIM, New York (2022)
3. Instituto Nacional de Estadística e Informática. Informe Técnico de la Producción Nacional mayo 2020–2021. https://www.inei.gob.pe/media/principales_indicadores/07-inf orme-tecnico-n07_produccion-nacional-may._2020.pdf. Accessed 10 Jul 2023
4. Ministerio de la producción, Estudio de Investigación Sectorial sector textil y confecciones 2020. https://ogeiee.produce.gob.pe/index.php/en/shortcode/oee-documentos-publicaciones/publicaciones-anuales/item/1065-estudio-de-investigacion-sectorial-sector-textil-y-confec ciones-2020. Accessed 10 Jul 2023
5. Instituto de Estudios Económicos y Sociales, Reporte Sectorial Industria Textil y Confecciones (2023). https://sni.org.pe/wp-content/uploads/2022/01/27-Industria-Textil-y-Confec ciones.pdf. Accessed 10 Jul 2023
6. Tapia-Cayetano, L., Barrientos-Ramos, N., Maradiegue-Tuesta, F., Raymundo, C.: Lean manufacturing model of waste reduction using standardized work to reduce the defect rate in textile MSEs. In: Proceedings of the 18th LACCEI International Multi-Conference for Engineering, Education, and Technology, pp. 1–8. LACCEI, Florida (2020)
7. Kumar, D., Mohan, G., Mohanasundaram, K.M.: Lean tool implementation in the garment industry. Fibres Text. Eastern Eur. **27**, 19–23 (2019)
8. Bukhsh, M., et al.: Productivity improvement in textile industry using lean manufacturing practices of 5S & single minute die exchange (SMED). In: Proceedings of the 11th Annual International Conference on Industrial Engineering and Operations Management, pp. 1–12. IEOM, Michigan (2021)
9. Cristobal, J., Quispe, G., Dominguez, F., Zapata, G., Raymundo, C.: Waste reduction with lean manufacturing model in an alpaca wool workshop. In: IOP Conference Series: Materials Science and Engineering, pp. 1–8. IOP Publishing, Bristol (2020)
10. Yesmin, T. and A. Zaheer: Productivity Improvement in Plastic Bag Manufacturing Through Lean Manufacturing Concepts: A Case Study. In: Applied Mechanics and Materials, pp.110–116. Trans Tech Publications, Switzerland (2011)
11. Alanya Veli, B.S., Dextre Vega, K.E.: Mejora del proceso de corte mediante la filosofía Lean Manufacturing en las MYPES exportadoras del Sector Textil. Universidad Peruana de Ciencias Aplicadas (UPC) (2020). http://hdl.handle.net/10757/653464
12. Condeso Carrizales, K.O., Nolasco Chuco, L.N., Salas Castro, R.F.: A combined model of lean manufacturing tools to increase efficiency in a peruvian textile company. In: Proceedings of the 8th International Conference on Industrial and Business Engineering, pp. 307–315. Association for Computing Machinery, New York (2022)
13. Poduval, P.S., Pramod, V.R.: Interpretive Structural Modeling (ISM) and its application in analyzing factors inhibiting implementation of Total Productive Maintenance (TPM). Int. J. Qual. Reliab. Manage. **32**(3), 308–331 (2015)
14. Bhamu, J., Sangwan, K.S.: Manufactura ajustada: revisión de literatura y problemas de investigación. Revista internacional de gestión de operaciones y producción **34**(7), 876–940 (2014)

15. Baptista, A., Abreu, L., Brito, E.: Application of lean tools case study in a textile company. Proc. Eng. Sci. **3**(1), 93–102 (2021)
16. Acosta-Ramirez, D., Herrera-Noel, A., Flores-Perez, A., Quiroz-Flores, J., Collao-Diaz, M.: Application of lean manufacturing tools under DMAIC approach to increase the NPS in a real estate company: a research in Peru. In: ACM International Conference Proceeding Series, pp. 70–76. Association for Computing Machinery, New York (2022)
17. Kara, E.: The effects of total quality management on employees' performance: a study on textile business. Bus. Manag. Stud. Int. J. **6**(3), 570–582 (2018)
18. Simegnaw Ahmmed, A., Ayele, M.: In-Depth analysis and defect reduction for ethiopian cotton spinning industry based on TQM approach. J. Eng. (U.K.) **34**, 1–8 (2020)
19. Zamora-Gonzales, S., Galvez-Bazalar, J., Quiroz-Flores, J.C.: A production management-based lean manufacturing model for removing waste and increasing productivity in the sewing area of a small textile company. In: Iano, Y., Saotome, O., Kemper, G., Mendes de Seixas, A.C., Gomes de Olivera, G. (eds.) Proceedings of the 6th Brazilian Technology Symposium (BTSym 2020), pp. 435–442 (2021). https://doi.org/10.1007/978-3-030-75680-2_49
20. Suryoputro, M.R., Sugarindra, M., Erfaisalsyah, H.: Quality control system using simple implementation of seven tools for batik textile manufacturing. In: IOP Conference Series: Materials Science and Engineering, pp. 1–10. IOP Publishing, Bristol (2017)
21. Verma, R.B., Jha, S.K.: Implementation of 5s framework and barriers modelling through interpretive structure modelling in a micro small medium enterprise. Int. J. Recent Technol. Eng. **8**(3), 7010–7019 (2019)
22. Neves, P., et al.: Implementing lean tools in the manufacturing process of trimmings products. In: Proceedings of the 28th International Conference on Flexible Automation and Intelligent Manufacturing (FAIM 2018), vol. 17, pp. 696–704. Elsevier BV, Netherlands (2018)
23. Ruiz, S., Simón, A., Sotelo, F., Raymundo, C.: Optimized plant distribution and 5S model that allows SMEs to increase productivity in textiles. In: Proceedings of the 17th LACCEI International Multi-Conference for Engineering, Education, and Technology, pp. 1–7. LACCEI, Florida (2019)
24. Aktar Demirtas, E., Gultekin, O.S., Uskup, C.: A case study for surgical mask production during the COVID-19 pandemic: continuous improvement with Kaizen and 5S applications. Int. J. Lean Six Sigma **14**(3), 679–703 (2023)
25. Mohan Prasad, M., Dhiyaneswari, J.M., Ridzwanul Jamaan, J., Mythreyan, S., Sutharsan, S.M.: A framework for lean manufacturing implementation in Indian textile industry. In: International Conference on Nanotechnology: Ideas, Innovation and Industries, pp. 2986–2995. Elsevier BV, Netherlands (2020)
26. Dudek-Burlikowska, M., Szewieczek, D.: The Poka-Yoke method as an improving quality tool of operations in the process. J. Achievements Mater. Manufact. Eng. **36**(1), 1–8 (2009)
27. Montalvo-Soto, J., Astorga-Bejarano, C., Salas-Castro, R., Macassi-Jauregui, I., Cardenas-Rengifo, L.: Reduction of order delivery time using an adapted model of warehouse management, SLP and Kanban applied in a textile micro and small business in Perú. In: LACCEI International Multi-Conference for Engineering, Education and Technology, pp. 1–8. LACCEI, Florida (2020)
28. Lista, A.P., Tortorella, G.L., Bouzona, M., Mostafad, S., Romeroe, D.: Lean layout design: a case study applied to the textile industry. In: Production, pp. 1–16. Scielo, Sao Paulo (2021)
29. Pacheco-Colcas, F.A., Medina-Torres, M.P., Quiroz-Flores, J.C.: Production model based on systematic layout planning and total productive maintenance to increase productivity in food manufacturing companies. In: ACM International Conference Proceeding Series, vol. 1, Issue 1, pp. 299–306. Association for Computing Machinery, New York (2022)
30. Rodriguez Concha, A., Rosadio Vela, J.R.: The implementation of lean manufacturing to increase productivity in a textile company: case of study. Thesis of Bachillerato, University of Lima, Lima (2022)

31. Quiroz-Flores, J.C., Collao-Díaz, M.F.: Application of lean manufacturing principles to increase productivity in SMEs manufacturers of baby clothes. In: Proceedings of the 2022 Congreso Internacional de Innovación y Tendencias en Ingeniería (CONIITI), pp. 1–5. IEEE, Bogotá (2022)
32. Arica-Hernandez, M., Llagas-Llontop, S., Khaburzaniya, I.: Implementation of lean manufacturing principles to increase productivity in SMEs in the manufacturing sector of clothing. In: Proceedings of the 2022 3rd International Conference on Industrial Engineering and Industrial Management, pp. 162–166. IEIM, New York (2022)

Lean Six Sigma and TPM for the Improvement of Equipment Maintenance Process in a Service Sector Company: A Case Study

Ana Laura Barriga⬥, Maria Gimena Gonzales⬥,
and Marcos Fernando Ruiz-Ruiz(✉) ⬥

Faculty of Engineering, Industrial Engineering Career, Universidad de Lima, Lima, Peru
20190221@aloe.ulima.edu.pe, {20180808,mruiz}@ulima.edu.pe

Abstract. Currently, competitiveness within the service industry has reached outstanding levels. In the pursuit of increasing the quality and efficiency of services, the approaches of continuous improvement acquire undeniable relevance. This research aimed to implement, analyze, and assess the effects of Lean Six Sigma (LSS) methodology within the service industry for process optimization through Total Productive Maintenance (TPM). The presented case study is a workshop for equipment maintenance where, through the DMAIC methodological design, Lean Six Sigma and TPM methodologies were implemented to increase process efficiency. The results show an efficiency increase of over 10%, reducing equipment DPMO by 53% and decreasing the total process time by 2 h. The value of the proposed model lies in its adaptability to companies providing technological equipment services in the regional context, as well as its contribution to recent studies on the application of TPM and LSS in the service sector.

Keywords: Lean Six Sigma · TPM · service industry · service sector · equipment maintenance · maintenance service · improvement process

1 Introduction

Facing increasing pressure to provide high-quality value, within a short time and at a minimal cost, organizations are confronted with a competitive reality that compels them to combine the various factors involved in service quality [13]. This competitiveness, along with the importance of the tertiary sector in the economy in recent years, requires a comprehensive management of the product or service lifecycle to be offered, including maintenance, repair, and revision of goods, services, and their components [17].

Given this scenario, organizations in the tertiary sector have been forced to seek continuous improvement in activities within their value chain through effective management methodologies and programs such as TPM [2]. TPM encompasses both the scheduling process and resource allocation to optimize asset utilization and maximize performance [3], while striving to maintain optimal conditions in equipment to avoid unexpected breakdowns, speed losses, and quality defects derived from the process [1]. On the other hand, Lean Six Sigma is an integration of Lean and Six Sigma methodologies, where

S.-H. Sheu (Ed.): IEIM 2024, CCIS 2070, pp. 155–169, 2024.
https://doi.org/10.1007/978-3-031-56373-7_13

the former focuses on waste reduction, time, and operating costs [6], and the latter on process optimization through variation reduction and customer needs focus [10]. Thus, Lean Six Sigma is considered one of the most effective techniques used by organizations in various economic sectors for process improvement, efficiency, and profitability increase [1].

In Peru, the increase in operating costs in maintenance and repair service organizations accounts for 21% due to service time delays [18]. Thus, the efficiency of the service sector in the country ranges from 30% to 60% solely in the city of Lima, being one of the lowest levels in Latin America [21]. In this context, the integration of methodologies for reducing delays and increasing performance and efficiency is essential to improve these ranges.

While the application of Lean Six Sigma in manufacturing and production environments is popular in research, its specific application in repair and maintenance services does not show the same progress [12]. Therefore, the aim is to contribute to the existing knowledge in the service sector, Lean Six Sigma, and TPM in Peru by generating an innovative proposal for the application of a new improvement model in companies in the sector, specifically in an equipment provision company in Arequipa, Perú. In this way, this model could be transferred to other contexts in the future and expand the vision of process improvement in Peru.

1.1 Lean Six Sigma and TPM Methodology

The focus of the service sector lies in the customer and their connection to the value proposition provided through this intangible benefit. Therefore, the development of innovative models for increasing service efficiency and quality becomes a priority for organizations [15]. In line with the pursuit of process improvement in speed and variability, the implementation of LSS increases customer satisfaction [7].

However, the competitiveness among different systems for service improvement leads to the search for additional strategies for maintaining quality. Thus, Total Productive Maintenance or TPM stands out as a strategy that links principles of engineering and total quality management [11]. Furthermore, TPM involves an innovative approach to equipment maintenance and efficiency improvement, breakdown elimination, and autonomous maintenance, representing a program capable of reducing equipment downtime and a philosophy that fosters a new attitude towards maintenance [23]. Both methodologies develop significant and notable improvement systems in various industry sectors. Therefore, the synergy between LSS and TPM, focused on the order fulfillment sector, allows for the establishment and consolidation of an efficient multifunctional team with a problem-solving mindset and a focus on maximizing efficiencies [20].

1.2 Studies on Lean Six Sigma and TPM in the Service Sector

Lean Six Sigma has been implemented in different companies in the manufacturing sector worldwide; however, the perceived benefits in these implementations can be replicated for the service sector [4]. Through various studies on Lean Six Sigma implementation [22], the main factors of variation in customer satisfaction and overall service efficiency were process standardization and waiting times. In addition, a banking service managed to increase customer satisfaction by reducing process defect variation and identifying valueless activities [8].

The fusion of Lean Six Sigma and TPM allows a significant reduction in costs, downtime, and an increase in quality standards and operational performance [13]. However, the implementation of these tools in the service sector presents certain complexities. Therefore, it is suggested to follow guidelines for Lean Six Sigma application, such as the proper management of critical success factors, quality characteristics, and performance indicators [16]. Despite that, complications arise when identifying difficulties in data collection due to non-automated interaction in services [14]. Additionally, these limitations converge with the high costs and organizational resistance that come with the Lean Six Sigma commitment to change [19].

2 Methodology

Given the nature of the service provided by the company, improvements can be generated both in the service provided and in the administrative process by making use of the eight pillars of TPM. Furthermore, the necessity of recognizing the main problems and measuring their impacts are aligned with the DMAIC method of Lean Six Sigma. Thus, to propose a model and validate an improvement in the productivity of maintenance services and order fulfillment, a case study was conducted on a company from Lima, Perú dedicated to the development, production, and commercialization of medical and safety application equipment. A pre and post-test with a mixed-method approach integrating qualitative and quantitative methods was applied, and a methodological design based on Lean Six Sigma methodology was established.

2.1 Case Study

The analysis focuses on a technology sector company located in Lima with over ten years of experience in the country. The company operates two authorized and equipped technical service workshops. This study specifically examines one of the workshops that recently started operations. The workshop employs a team of two operators per day, working in two shifts. The workshop spans 48 m^2 and has a fleet of 125 gas detectors for each shift, 250 units in total. Before delivery, every unit undergoes a bumptest and a full charge to ensure operability.

2.2 Methodological Flow

Figure 1 presents the methodological flow of the research and the tools employed in each phase.

As shown, Lean Six Sigma methodology, employing the DMAIC continuous improvement steps, was integrated with Total Productive Maintenance (TPM), applied in the measurement, analysis, and improvement phases. Also, different Lean tools were applied in each stage for a more detailed case study.

Firstly, in the Define step, it is necessary to understand the whole process, how the workshop works, and identify what is failing in the equipment. While maintaining the

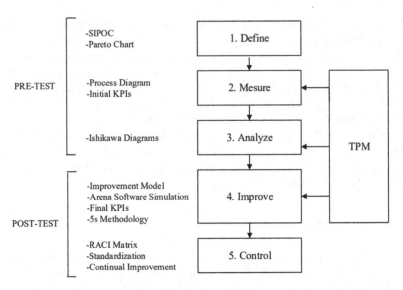

Fig. 1. Methodological flow

focus of both methodologies, a SIPOC diagram was used to observe the whole process and understand its overall operation. The most common equipment failures were also identified using a Pareto diagram. Subsequently, Measure step is developed to achieve a detailed understanding of the maintenance process and visualize the times for each activity. A Process Diagram of the initial scenario (pre-test) was utilized. Also, to measure the improvement, the indicators were selected considering TPM concepts to engage the management of equipment as shown in the Table 1. These indicators were calculated based on the initial scenario.

Table 1. Indicators

Indicator	Formula
Process Efficiency	$\frac{Standardlaborhours}{Workedhours} \times 100\%$
Overall Equipment Effectiveness	$Availability \times Performance \times Quality$
Defect Per Million of Opportunities	$\frac{TotalDefects}{TotalOpportunities} \times 100\%$

Subsequently, an Ishikawa diagram was developed to analyze the root causes of the object of study. Additionally, for the Improvement step, Lean Six Sigma tools were utilized as 5s Methodology. The construction of the improvement model began for subsequent simulation using Arena 16.0 Software. This allowed to obtain the results of the model (post-test) without intervening in the company. Finally, the KPIs were measured using the proposed model. For the Control step, a RACI matrix, standardization, and continuous improvement were proposed.

It is important to mention that throughout the research process, continuous validations were conducted with three experts in Lean Six Sigma, TPM and the service sector. Additionally, four advisory members from the company corroborated the application in each predetermined phase.

3 Results

The results follow the DMAIC process presented in the methodological flow.

3.1 Process Definition

Firstly, a SIPOC diagram was employed to represent and understand the maintenance process of equipment in general terms. The process begins when the user approaches the workshop and hands over the equipment to the service engineer (SE). At that moment, if the user encountered any issues with the equipment during use, they document it on a form. The SE receives the equipment and performs maintenance (Table 2).

Table 2. SIPOC of the process

Suppliers	Inputs	Process	Outputs	Customers
User 1	Used equipment	Complete the Failure Report	Failure Report Format complete	Service Engineer
	Failure Report Format			
Service Engineer	Failure Report Format complete	Equipment maintenance	Operational repaired equipment	User 2
	Used equipment			

The Failure Report filled out by users facilitates the identification of Critical-to-Quality (CTQ) characteristics. The following failures were identified: equipment shutdown (battery failure), inoperable buttons (casing failure), measurement error (sensor

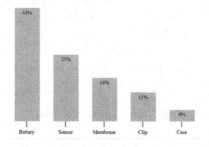

Fig. 2. Pareto Chart of Equipment Failures

failure), perforated sound membrane (membrane failure), loose fastening clip (clip failure). To visualize the frequency of occurrence of these failures, a Pareto diagram is
shown in Fig. 2.

It is observed that 80% of failures are concentrated in the battery, sensor, and membrane. It is worth mentioning that these failures are addressed by replacing the faulty
component with a new one (Table 3).

Table 3. Repair Times by Identified Failure

# Failure	Failure Type	Repair Time
1	Battery	20
2	Sensor	40
3	Membrane	80
4	Clip	5
5	Case	20

3.2 Process Measurement

A process diagram with times was utilized to understand the maintenance service process
for the equipment (Fig. 3).

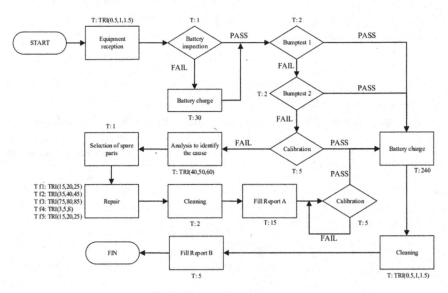

Fig. 3. Process Diagram (min).

The process starts with equipment reception and a quick visual inspection by the
SE to promptly check if the equipment has over 50% battery charge. If it fails to meet

the criteria, it is moved to the charging area. Subsequently, the equipment undergoes an initial bumptest, which tests the operability of batteries, alarms, and sensors. The bumptest equipment can accommodate 10 units at a time. If the test is successful, the equipment is moved to the charging area. If any equipment fails the test, a repeat test is performed only for the failed units. If it passes, it is placed in its charging station. When the equipment fails, it indicates a potential deviation, so a calibration is performed in the same bumptest equipment. If it fails, it means that the equipment requires a component replacement. In this case, the engineer conducts an inspection to identify the root cause of the failure to select and transfer the replacement component.

After the repair, the equipment is cleaned and goes through calibration again. When it passes this adjustment, it is considered operational and is moved to its charging station. Also, SE must document this corrective maintenance in a detailed report. Once the equipment completes its 4-h charging period, it is transferred to the workbench for final cleaning and remains there until delivery.

Once the process has been understood, it is necessary to calculate the previously established KPIs to evaluate the current situation:

– Process Efficiency: The workshop operates with two 8-h shifts: day shift and night shift, totaling 16 h of working time. However, the average time taken by employees to complete the target of inspected equipment is 18 h. Therefore, the efficiency percentage in the process is calculated.

$$Efficiency = \frac{Standard labor hours}{Worked hours} \times 100\% = \frac{16}{18} \times 100\% = 89\%$$

– Equipment Efficiency (OEE): On average, the workshop can complete maintenance on 250 equipment units daily. However, records show that the number of defective units per day averages 15. Therefore, the quality rate (Q), given by the percentage of functioning equipment out of the total received, can be calculated as 94%. Additionally, considering that the workshop operates for 30 days with 2 scheduled rest days, resulting in 28 days of planned maintenance and operation, the availability (A) is calculated as 89%. The average performance rate in the workshop is calculated monthly, and the records of the last 3 months indicate a performance (PE) of 85%.

$$OEE = Availability \times Performance \times Quality = 0.89 \times 0.85 \times 0.94 = 0.71 = 71\%$$

– Defects Per Million Opportunities (DPMO): Based on the workshop reports from the past three months, a sample of 250 repaired equipment units is considered daily, of which 15 units have defects. Furthermore, the opportunity for defects represents the 5 identified failures during equipment inspection. Accordingly, it is calculated that there can be 12,000 defective units per million inspected units. This result represents a 50% increase in defective units compared to the workshop's initial maintenance facility.

$$DPMO = \frac{Total Defects}{Total Opportunities} = \frac{15 \times 1000000}{250 \times 5} = 12000$$

3.3 Process Analysis

To identify the root causes of low performance in the maintenance process within the workshop, a cause-and-effect diagram focused on four main factors was created: environment, operator, materials, and method (Fig. 4).

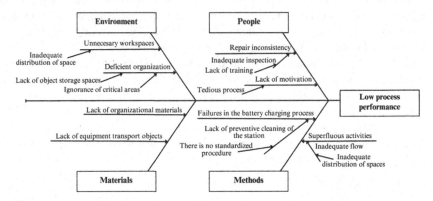

Fig. 4. Ishikawa Diagram - Low process efficiency

The presented diagram reveals fundamental root causes whose elimination would significantly reduce rework and process times. The criteria were selected by observing and evaluating the entire process. From the analysis of the diagram, it is concluded that a new space layout with standardized workflows and procedures is necessary. Additionally, the entire workshop must be organized.

Similarly, to identify the root causes of equipment failures, an Ishikawa diagram was employed focused on the five previously identified failures with the Pareto Chart (case, membrane, sensor, clip, and battery). This second analysis is important to reduce the DPMO indicator value and improve the Equipment efficiency. The Fig. 5 presents the root causes of the identified problem.

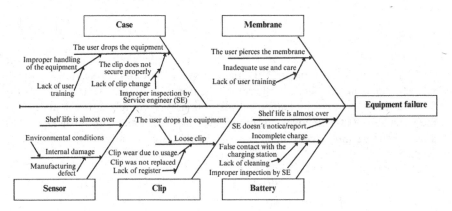

Fig. 5. Ishikawa Diagram - Equipment failures

It is concluded people and their handling of the equipment significantly influence their failure. The service engineer must receive training for the proper repair and calibration of the equipment, and users should also receive training on the correct use and handling of the equipment. Additionally, there is a need to track the lifespan of the sensors and batteries.

3.4 Proposed Improvement

In order to develop an improvement proposal for the maintenance process, several TPM tools have been selected to improve both the efficiency of the maintenance process and reduce equipment failures. The 5S Program, also known as Kaizen, has been chosen, aiming to reduce the risk of defective equipment and reduce the process time (Table 4).

Table 4. 5s Phases.

5s Phases	Proposal
Seiri-Sort	Discard used spare parts. Selection and relocation of elements of different uses that were piled up. Reduction of inventories of cleaning supplies
Seiton – Set in Order	Add a floating shelf above the maintenance area for the spare parts and classify them with organizers considering the frequency of use. Add color-coded boxes for each A redesign of the flow was made to reduce distances and the total process time. For this, a new space distribution is proposed to allow a continuous process flow
	Add a basket for equipment transportation in the workshop
	Define areas for spare parts, office supplies, cleaning supplies and stored items
	Delimit the distinct areas in the floor
	Reorganize workbenches for better alignment with the process
Seiso-Shine	Remove empty boxes in the workshop. Generate a clean and tidy workplace
	Implement an inspection and cleaning checklist. Establish cleaning schedules
Seiketsus-Standarize	Implement visual controls for monitoring and tracking daily objectives Standardize the flow and the service process by establishing it within protocols and manuals

(*continued*)

Table 4. (*continued*)

5s Phases	Proposal
Shitsukes-Sustain	Complete a weekly 5S evaluation form and track the compliance percentage Stablish coordination meetings to identify opportunities for continuous improvement Deploy a 5s panel which includes photographic records, tracking forms, and the cleaning plan Implement a program of recognitions for continuous improvement and its implementation

Additionally, other TPM pillars have been considered for the improvement model:

– Jishu Hozen: Also known as autonomous maintenance, operators are responsible for keeping all the equipment they use in the maintenance in excellent conditions to prevent deterioration. They must also ensure that calibration certificates are always up to date by stablishing a preventive calibration plan.
– Kobetsu Kaizen: involves small improvements and is carried out by involving people at all levels of the organization. A meticulous procedure is followed to systematically eliminate losses using tools as Poka Yoke devices. Sequential Poka Yokes will be used with the new proposed process to maintain the order. Information Poka Yokes will also be employed: graphic instructions for the bump test and calibration process. Also, a new layout is proposed as shown in Fig. 6. The proposed layout is designed to minimize to minimize transfer time and enable a continuous process flow.

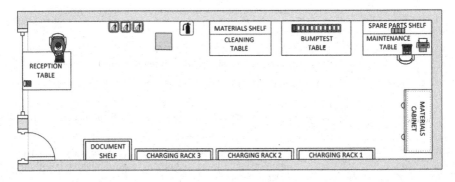

Fig. 6. Proposed Layout

– Education and training: TPM education and training programs have been prepared to achieve two objectives. First, managers will learn to effectively plan and control the maintenance process, suggesting timely changes in the event of deviations, breakdowns, or defects. Second, SE will study the basic principles and techniques of maintenance and calibration adjustments. They will also be trained in equipment technology and develop specialized maintenance skills. They will learn to identify process deviations.

Once the improvements have been identified, a comparison between scenarios was performed by constructing the current process flow and the proposed process flow. The proposed process included the whole improvement previously developed and was reflected in the new workflow and the reduction of transportation, inspection, and maintenance times, considering that the employees were well-trained, and equipment errors reduced by almost 50% as 5S system effectively reduces the errors made by workers through education and training [4]. In this way, the process was portrayed in detail by simulating the model generated from the proposed improvements using Lean and TPM tools through Arena 16.0 software. The controlled variables considered were hours of work, number of equipments and number of operators. On the other hand, uncontrolled variables were time of the Process and number of failures. The simulation was performed for an 8-h shift, where an operator receives a maximum of 125 equipment units for inspection (Fig. 7).

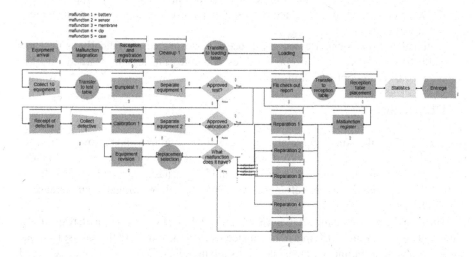

Fig. 7. Proposed Simulation Model

The simulation was successfully developed as the results shown a reduction of times in almost every step and a reduction of failures from 15 to 8 per day. For scenario comparison, the final KPIs were obtained by the simulation results. Table 5 shows the obtained indicators. As observed, the initially established indicators in the research have been improved with the new established model. Thus, by reducing times, streamlining processes, and implementing a new process flow, the overall efficiency of the process and equipment were improved.

Table 5. Styles available in the Word template

Indicator	Current Scenario	Proposed Scenario
Process Efficiency	89%	94%
OEE	71%	87%
DPMO	12000	6400

3.5 Results Control

The control phase includes the following activities:

- Communication management: Effective communication is crucial for the successful implementation of any improvement methodology, as it supports change. To ensure effective communication, a Responsibility Assignment Matrix (RACI Matrix) was utilized.
- Standardization: includes generating formats for the implementation of improvements within the equipment maintenance process. To achieve this, a progress management format was considered.
- Continual Improvement: Review the process performance periodically against the established metrics. Identify opportunities and integrate lessons learned into the process for further optimization.

4 Discussion

After conducting the study and analyzing the results obtained, an improvement in all proposed indicators can be observed. Furthermore, the technical gap in the country, corresponding to an efficiency between 32.7% and 58.1%, has been overcome. As shown in Table 5, the achieved process efficiency was 94%, which in turn led to an increase of OEE to 87%.

Through the presented model, Lean Six Sigma and TPM were integrated, resulting in improved efficiency and OEE in a maintenance workshop, similar to the case study conducted on a vehicle engine maintenance workshop in India in 2017, where they achieved a 13% increase in OEE and a 7% increase in process efficiency [3]. While implementing Lean Six Sigma in services is more complex compared to its implementation in the manufacturing sector [7], the study's results showed that an appropriate implementation model adapted to services facilitates the application of this methodology.

Furthermore, the use of TPM in maintenance services allows for maintaining users' equipment in optimal conditions, guaranteeing an efficient and high-quality service [11]. Also, TPM focuses on maximizing OEE factors such as availability, quality, and performance [5]. It is important to mention that the intangible nature of services limits the applicability and effectiveness of TPM [10]; however, several authors affirm that the Overall Equipment Effectiveness (OEE) framework has become a leading measure for performance improvement by identifying areas of loss in the process [9]. This was evidenced in the case study, as the application of OEE allowed to focus on equipment failures and identify the cleaning and repair areas as critical for the overall time.

As a result, it is concluded that maintenance management must prioritize the training of service engineers and the training provided to equipment users since the primary causes of failures were linked to inadequate usage and maintenance. In addition, the pillars of TPM, such as autonomous and preventive maintenance, should be implemented. Still, their application may require both time and financial investment, as TPM needs to be embraced as a cultural framework, as indicated in other studies [5].

The Lean Six Sigma methodology has a fundamental focus on waste elimination and process optimization, enabling the identification and elimination of non-value-added activities [3]. In the presented case study, repetitive processes such as cleaning were eliminated, and time was reduced by proposing a more suitable space distribution aligned with the process flow and the implementation of the 5S. While a new workflow was established, it was constrained by the existing infrastructure. Furthermore, the workshop's management insight must support and approve the improvement program.

Although additional effort is required, Lean Six Sigma and TPM methodologies enable a greater optimization of processes and equipment, leading to significant improvements in physical assets and service efficiency. This was evidenced by the conducted simulation: the achieved standard in the work represents the effectiveness of the tools in the equipment maintenance process, reaching an efficiency higher than 40% of the industry standard.

5   Conclusions

Based on the obtained results, it is demonstrated that the implementation of Lean Six Sigma and TPM methodologies significantly increases process efficiency in service sector companies. By constructing a solid framework for problem identification and resolution, process optimization, and equipment quality maximization, the efficiency indicator was improved from 89% to 94%.

While the research presented a model for improving the performance of an equipment order fulfillment process, it is crucial to continue investigating adaptable and easily applicable Lean Six Sigma tools for the service sector. Given that the proposed model based on Lean Six Sigma and TPM provides evidence of effectiveness in improving process performance and reducing defects for tertiary sector companies, it can be validated that the combination of Lean Six Sigma principles and TPM pillars allows for the construction of a comprehensive and detailed methodology for eliminating idle time, superfluous subprocesses, and reducing defects to optimize the equipment order fulfillment process. However, it is important to highlight that the proper implementation of these tools requires a deep understanding of the tangible and intangible characteristics of the service model being improved, as well as constant reinforcement through training for the workforce.

By exploring TPM from a services perspective, doors are opened to new approaches beyond equipment maintenance. This delves into the maintenance of functional systems and assets, automation, and the importance of personnel training and development in equipment maintenance and operation. Based on this, further in-depth research is recommended regarding the necessary continuity for TPM maintenance in a service improvement model to gain a better understanding of the integration of maintenance and service efficiency. Additionally, it is recommended that future studies explore the combination of Lean Six Sigma and TPM in different subsectors of the service industry, conducting comparisons that provide a comprehensive perspective on the different tools and approaches to be adapted within them.

References

1. Alblooshi, M., Shamsuzzaman, M., Karim, A., Haridy, S., Shamsuzzoha, A., Badar, M.A.: Development of a framework for utilising lean six Sigma's intangible impacts in creating organisational innovation climate. Int. J. Lean Six Sigma **14**(2), 397–428 (2023). https://doi.org/10.1108/IJLSS-08-2020-0117
2. Al-refaie, A., Lepkova, N., Camlibel, M.E.: The relationships between the pillars of TPM and TQM and manufacturing performance using structural equation modeling. Sustainability (Switz.) **14**(3) (2022). https://doi.org/10.3390/su14031497
3. Alsubaie, B., Yang, Q.: Maintenance process improvement model by integrating LSS and TPM for service organisations (2017). https://doi.org/10.1007/978-3-319-62274-3_2
4. Aminudin, O., Zainol, M.: Implementation of six sigma in service industry. J. Qual. Measur. Anal. **10**(2), 77–86 (2014)
5. Annamalai, S., Suresh, D.: Implementation of total productive maintenance for overall equipment effectiveness improvement in machine shop. Int. J. Recent Technol. Eng. **8**(3), 7686–7691 (2019). https://doi.org/10.35940/ijrte.C6212.098319
6. Bader, B.H., Badar, M.A., Rodchua, S., McLeod, A.: A study of the balancing of lean thinking and stakeholder salience in decision-making. The TQM J. **32**(3), 441–460 (2020)
7. Caiado, R., Nascimento, D., Quelhas, O., Tortorella, G., Rangel, L.: Towards sustainability through green, lean and six sigma integration at service industry: review and framework. Technol. Econ. Dev. Econ. **24**(4), 1659–1678 (2018). https://doi.org/10.3846/tede.2018.311 910.3846/tede.2018.3119
8. Chakrabarty, A., Tan, K.: The current state of six sigma application in services. Managing Serv. Qual. **17**, 194–208 (2007)
9. Facchinetti, T., Citterio, G.: Application of the overall equipment effectiveness to a service company. IEEE Access **10**, 106613–106640 (2022). https://doi.org/10.1109/ACCESS.2022. 3211266
10. Ganjavi, N., Fazlollahtabar, H.: Integrated sustainable production value measurement model based on lean and six sigma in industry 4.0 context. IEEE Trans. Eng. Manage. **70**(6), 2320–2333 (2023). https://doi.org/10.1109/TEM.2021.3078169
11. Hardt, F., Kotyrba, M., Volna, E., Jarusek, R.: Innovative approach to preventive maintenance of production equipment based on a modified TPM methodology for industry 4.0. Appl. Sci. (Switz.) **11**(15) (2021). https://doi.org/10.3390/app11156953
12. Hill, J., Thomas, A.J., Mason-Jones, R.K., El-Kateb, S.: The implementation of a lean six sigma framework to enhance operational performance in an MRO facility. Prod. Manuf. Res. **6**(1), 26–48 (2018). https://doi.org/10.1080/21693277.2017.1417179
13. Kumar, M., Antony, J., Singh, R.K., Tiwari, M.K., Perry, D.: Implementing the lean sigma framework in an Indian SME: a case study. Prod. Planning Control **17**(4), 407–423 (2006). https://doi.org/10.1080/09537280500483350

14. Legman, I., Rozalia, M., Kardos, M.: An innovative tool to measure employee performance through customer satisfaction: pilot research using eWOM, VR, and AR technologies. Electronics **12**, 1158 (2023). https://doi.org/10.3390/electronics12051158
15. Miremadi, M., Goudarzi, K.: Developing an innovative business model for hospital services in iran: a case study of Moheb hospitals. Leadersh. Health Serv. **32**(1), 129–147 (2019). https://doi.org/10.1108/LHS-10-2017-0063
16. Owee, T., Peidi, L., Kim, L., Choon, O., Chin, G., Kadir, B.: Critical success factors for lean six sigma in business school: a view from the lecturers. Int. J. Eval. Res. Educ. (IJERE) **11**(1), 280–289 (2022). https://doi.org/10.11591/ijere.v1lil.21813
17. Pezzotta, G., et al.: The product service system lean design methodology (PSSLDM): integrating product and service components along the whole PSS lifecycle. J. Manuf. Technol. Manag. **29**(8), 1270–1295 (2018). https://doi.org/10.1108/JMTM-06-2017-0132
18. Pinedo-Rodriguez, K., Trujillo-Carrasco, L., Cabel-Pozo, J., Raymundo, C.: Resource management model to reduce maintenance service times for SMEs in lima-Peru (2022). https://doi.org/10.1007/978-3-030-85540-6_148
19. Rishi, J., Srinivas, T., Ramachandra, C. BC, A.: Implementing the lean framework in a small & medium & enterprise (SME) - a case study in printing press. IOP Conf. Ser. Mater. Sci. Eng. (376) (2018). https://doi.org/10.1088/1757-899X/376/1/012126
20. Syaifoelida, F., Amin, I., Megat Hamdan, A.M.: The designing analysis process of constituent attributes by using VSM and six sigma to enhance the productivity in industry of bearings. IOP Conf. Ser. Mater. Sci. Eng. **788**(1) (2020).https://doi.org/10.1088/1757-899X/788/1/012021
21. Tello, M.: Índice de eficiencia técnica de las empresas de Perú. Desarrollo y Sociedad **1**(90), 111–151 (2022). https://doi.org/10.13043/dys.90.4
22. Valdivia, A., Villavicencio, J., Chavez, R., Collao, M.: Service model under the lean service and machine learning approach to increase external user satisfaction: a case study in the health sector SMEs in Peru. ICIBE **2022**, 27–29 (2022)
23. Xu, M., Gao, X., Zhang, F.: An approach of implementing SW-TPM in real-time operating system (2019). https://doi.org/10.1007/978-981-13-5913-2_7

Service Model Based on Lean Service and Agile Methodology to Increase the NPS Index in a Company in the Security Sector

Christian Moscoso-Zuñe [ID], Vanessa Zuloaga-Luna [ID], Martin Collao-Diaz[✉] [ID], and Eduardo Del Solar-Vergara [ID]

Facultad de Ingeniería, Universidad de Lima, Lima 15023, Perú
{20130863,20163759}@aloe.ulima.edu.pe, {mcollao, edelsol}@ulima.edu.pe

Abstract. The contractual relationship of a security company is dependent on the subjectivity of the owner of the service and variable to the daily incidents of the service. The NPS helps companies in this sector to quantify the ailments of the service and captures them as points of improvement. In this line, the security company has in the first place the management of its invoices (35%) and in the following line the management and operational support (32%). Impacting the company with losses of close to one million dollars. To alleviate this ailment, agile methodologies and lean service were applied, managing to reduce customer perception of invoice management by 57%, and a 30% reduction in operational perception. Impacting three hundred thousand dollars.

Keywords: Lean service · Agile Methodologies · Personal security

1 Introduction

Currently, service companies play a fundamental role in the economic development of each country. According to a report from the INEI, 13% of employment generation in Peru corresponds to this sector and absorbs 40.5% of the employed population, being above agriculture, fishing, and mining. However, each business model has points of improvement that must be identified, analyzed, and resolved. In the security sector, two of the main problems to improve are: the elevated level of staff turnover and the low level of customer satisfaction. The latter is, since being an intangible value, the customer's rating is measured by various subjective indicators. To understand this index, it is important to specify that the "Net Promoter Score" focuses on evaluating the quality of service perceived by the client, through a scoring question. The same one that categorizes the responses into three groups: detractors, passives, and promoters.

Specifically, in the asset security analysis company, the main reasons for the problem are two: inadequate management of invoice issues and a low level of operational management. This problem has also been identified in other sectors. For example, in the hotel industry, where the relationship between customer satisfaction and retention was

S.-H. Sheu (Ed.): IEIM 2024, CCIS 2070, pp. 170–182, 2024.
https://doi.org/10.1007/978-3-031-56373-7_14

analyzed, emphasizing three main pillars to increase NPS. The first seeks to ensure that all the staff involved behave in the best way with the client through training. The second, identify what criteria the client values. And the third, resorts to work standardization to align all the functions of collaborators (Barusman and Rulian, 2021). Under this scenario and focusing on the two main reasons that afflict this company, four root causes were identified collected from the company's statistical information in the last three years. The fundamental causes of the first reason are: error in the content of the invoices (58%) and failure to comply with the agreed date (42%). While the fundamental causes of the second reason are the following: non-existent control and supervision plan (61%); and inefficient contingency plan before the implementation of the checkpoint (39%).

Within this problem, which brings the company losses of around four million soles a year. Two great tools have been identified as palliatives to the problem studied: agile methodologies and lean service. Although there is not enough documentation on the good management of "NPS" in the sector, initiatives have been collected in similar sectors. For example, in the financial and security sector there was a 45% increase in complaints compared to 2020, linked to customer service. Along these lines, the measurement of time was identified as the main ally to contextualize the situation. With this, the financial company used a continuous improvement model using lean service, which reduced the variability of the customer service process. This execution benefited the company with an 8% increase in its service level, reduced waiting time by 73% and decreased operation time by 35% (Viacava, 2021). In another case, agile methodologies were implemented in a technological solutions development company whose objective is to conduct 100% of the deliveries according to an established schedule. For this, the "SCRUM" methodology was determined as the most appropriate tool. With which a diagnosis, change planning, implementation, control, and evaluation were conducted. Likewise, it incorporates the client in each small project, to strengthen client satisfaction. The results of this study are increased sales, improved delivery times and product quality. All of this aimed at recovering customer satisfaction, which the company said had been lost in 2019 (Baldoceda, 2021).

As mentioned in previous lines, it is especially important to conduct this research, not only because there are currently few related articles, but also because of the level of relevance that service companies are having in Latin America.

To publicize the proposal, this academic article is divided into the following sections: State of the Art, which details the background of the problem from different approaches of the authors; Contribution, where the theoretical foundation of the proposed models is explained, as well as the stages that compose them and their respective indicators; Validation, where the results of the pilot tests are described; Discussion; Conclusions; and References.

2 Literature Review

2.1 Standardization of Procedures with Lean Service

Lean procedures have been applied in various business models, such as services [1, 2]. Although it has been applied for a longer time in industries, the service category is not exempt from inefficiency problems that impact customer satisfaction [3]. For example, non-compliance with orders on agreed dates generates cost overruns related to a contingency plan to reverse the failure.

Beyond the type of Lean, they all have the level of customer satisfaction as an indicator, which is affected by the fulfillment of the user's expectation based on the quality of service, in the case of the service sector, by the quality of care of the staff that provides it. Therefore, decent work management based on the human group that is the first line of service, guarantees the superior performance or success of the company's objectives and projects [4].

Today corporations have set their sights on tools and methodologies that help identify the landscape, as well as solutions to improve quality at all levels of service, always with an efficient approach. In this line, the Lean Service methodology has helped reduce time and costs, increasing efficiency in favor of standardized processes. In other words, a more productive company, without high percentages of losses and under strategies designed for cost and corporate impact. This applied solution has obtained various achievements, for example, in the medical services industry, it was possible to reduce delays in operating rooms [5]. In addition, it evidenced the advantages of standardization with the development of processes of daily functions of the medical team.

2.2 Agile Management Model

A perfect complement, which different corporations have incorporated along with Lean, are agile methodologies [1]. They allow improving project management, favoring the performance of the company and the development of its employees. One of the aspects that these methodologies work on is the concept of teamwork based on an effective use of resources and time spent. Thus, creating the concept of multifunctional team, the same one that enhances the management of processes based on effective communication, a learning and feedback approach, and the integration of the team, where each one contributes based on their specialization [6, 7]. All this under an orderly scheme thought of the business objective and with schedules grounded to impact at the best moment [8].

The scrum methodology allows for a better dimensioning of projects, correctly estimating delivery dates and favorable feedback for team members. Its implementation in a medical department has achieved efficient management in terms of processes and improving the service offered to customers [1]. In addition, it achieved corporate sustainability in an Indian manufacturing company. Improving the response time and level of service offered [7].

2.3 Methodologies that Improve the Factors Related to the Level of Satisfaction

The net promoter score (NPS) estimates customer satisfaction, measuring the percentage of promoters and detractors [9]. To obtain a better result of this indicator, it is essential to reduce detractors and increase promoters [10]. Aspects such as order delays and long waiting times are those that increase the percentage of detractors. Claims for service, product, attention and even price also positively or negatively affect customer satisfaction [11].

The Lean Service methodology improves customer satisfaction. Since, by standardizing processes, it is possible to reduce times, improve service and reduce production costs [10, 12]. On the other hand, the agile Scrum methodology allows us to manage a work team in the best way. Establishing a structure for the teams involved in projects and the fulfillment of their objectives. The success of the development of this agile methodology strongly depends on the maturity of the work team [13].

2.4 Improvement of the NPS Implementing Agile Methodology and Lean Service

Customer satisfaction is a truly relevant indicator to measure the performance of a company. Since, this indicates the percentage of satisfied customers; and opportunities for improvement in processes that afflict customers and are potential nuisances for other customers.

Standardization helps to parameterize these processes and evidences the improvement by making them more efficient [12]. As well as to restructure a process flow, increasing, on the one hand, the production capacity, and on the other, reducing the delivery time of orders. This was possible by eliminating downtime, making a correct programming of activities and relocation of resources [14].

There are various methodologies to increase customer satisfaction. Among them are Scrum and Lean Service. Scrum allows us to make teamwork more effective. Achieving the fulfillment of objectives in an organized and methodical way [1, 7]. On the other hand, Lean Service focuses on the efficiency of processes. Eliminating dead times, parameterizing them, implementing indicators to measure and control their performance [12, 15].

3 Contribution

3.1 Implementation of the Agile Scrum Methodology

The Scrum methodology connects with two fundamental causes of non-compliance with the agreed date and an inefficient contingency plan for the implementation of checkpoints. The application of this methodology seeks to reduce the percentage of non-compliance with agreed dates and improve the contingency plan to cover missing positions, all in favor of improving the quality of service. The greater the supervision, the greater the customer satisfaction. The following Figs. 1, 2 and 3 present the Method and Model proposal implementing the Scrum methodology for the company.

3.2 5S Implementation

To improve the supervision plan, it is necessary to project the field of work, therefore, in cooperation with the service company, there was a field visit where the agents involved in to identify the supervision process. Although the supervisor is the main executor of this work, there are relevant agents in the management of this main activity that directly affects the NPS. During the visit, it was possible to identify that the head of base, the operational assistant manager, the dispatch coordinator, and the operational administrator are influential agents in the process and are primarily responsible for its effectiveness.

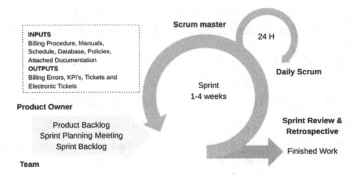

Fig. 1. Proposed model for the Scrum

As part of the management of the visits included in the schedule, a time record was kept for the supervisors of the four Lima bases: central base (Lince), south base (Chorrillos), north base (Callao) and east base (Ate).With this information it was possible to categorize the times that the supervisor develops in his day to day, identifying that only 13.9% supervise and there is a considerable percentage of development of administrative activities.

For the development of the 5S tool, the organization will begin. For this, a summary of all the procedures executed by the supervisor will be managed and categorized according to their execution time in month. According to the operations manual managed by the company, the supervisor manages about fifteen procedures with about thirty tasks. After having the activities hierarchical, the first 3 will be taken to continue with the process and the continuity of the last 5 will be evaluated, in favor of joining it with other activities or seeing their elimination from the supervisor's functions, either due to lack of relationship with the profile or for lack of contribution to the value chain.

With the three main ones, we will proceed with the order, analyzing the procedure in general and proposing a new structure, promoting the efficient use of time. Always in favor of cleaning the structure, respecting the new steps, and managing the monitoring for proper follow-up. To this end, unplanned visits will be established in 2 weeks to validate the discipline of correct follow-up.

The following figures present the Method and Model proposed for the company using Lean Service.

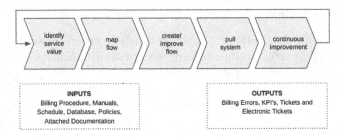

Fig. 2. Proposed model for the Lean Service

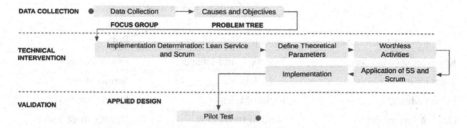

Fig. 3. Proposed design

Expected results-metrics. Regarding the indicators, it was determined based on the proposed integrating design, which consists of three pillars: Billing error rate, NPS and 5S Audit. These will be guiding pillars to identify the correct implementation management.

Billing error rate (F). It is the percentage of erroneous invoices out of the total number of invoices issued.

$$F = \text{Number of wrong invoices/Number of invoices issued} * 100 \tag{1}$$

Net Promoter Score (NPS). It is a metric used to measure the level of customer satisfaction.

$$NPS = \% \text{ promoters} - \% \text{ detractors} \tag{2}$$

5S Audit. It is the total average of the five S's.

$$5SC = \left(\sum (\text{Organize} + \text{Sort} + \text{Clean} + \text{Standardize} + \text{Discipline})\right)/5 \tag{3}$$

4 Validation

4.1 Case Study Model Contact

Organizational communication is an important and necessary factor for the implementation of a pilot, in which the reception of potential improvements in each area is identified, as well as the previous knowledge that the team has about the tools to be implemented. Given this, and with the correct support of senior management, it was identified that eight out of ten people surveyed had the basic knowledge for development.

4.2 Case Study Model Management

As can be seen in Tables 1, 2 and 3, it is necessary to assemble a work team consisting of the product owner, scrum master and developers.

With a focus group, it was identified that supervisors do not prioritize supervision as such, their focus is concentrated on administrative tasks that are not necessarily part of their functions as a position, but rather responsibilities as direct boss of a group of PVS

Table 1. Operational and billing scrum work team

Work position operational	Work position billing	Role
Key Account Manager	Key Account Manager	Product Owner
Team ULima	Team ULima	Scrum Master
Project Manager	Project Manager	Development Team
Head of Innovation	Head of Innovation	Development Team
Operations Manager	Billing analyst	Development Team
Operations Assistant Manager	Finance Assistant Manager	Development Team
Head of Operational Base	Commercial analyst	Development Team
Suffers RRHH	Legal assistant	Development Team
Planning Coordinator		Development Team

(private security agents). There were surprise reactions to the debate on the subject. The lack of recognition of the supervisor was also touched on, as well as the devaluation of the "checkpoint" position, one that instead of being sold as the position that covers the position due to lack, is identified as the one that "is at the client for a few days and you do not need to familiarize yourself with the regulations." Finally, there is no clear supervision manual that is used as the standard for it.

Thanks to a measurement of H-H, it was identified that only 13.9% supervise and there is a considerable percentage of development of administrative activities (32%). This last point was validated in the field meeting, where the collaborators were aware that as administrative improvements are applied, support or follow-up management was charged to the supervisor's activities, therefore, it is important to execute the 5S tool in field and apply it in favor of supervision, aiming at quality supervision.

On the billing side, the sale at the amount level rather than the amount of customer was the priority to execute the organization of the classes. Likewise, a structure was managed based on the company's Pareto, the same ones that have the largest participation in the company and as such increase their level of importance due to the degree of affectation. The same one that graduates or qualifies the client in three levels related to their payment behavior.

Table 2. Classification of customers according to the percentage of total sales

Class	Total customers	Sale (millions)	%
A	42	17.84	75.1%
B	64	3.53	14.8%
C	204	2.38	10.1%
Total	310	23.75	100%

Table 3. Type of customers according to payment behavior

Type	Description
1	Customer who has debt and/or provision up to date
2	Client who has a debt and/or overdue provision of one month
3	Customer who has a debt and/or overdue provision of two months or more

Crossing the class with the type gives rise to the action plan for each situation. To maintain an orderly and playful categorization, a traffic light was applied where the green color reflects follow-up and basic review actions, yellow, actions with the involvement of the commercial advisor in charge, and red, action plans oriented to company decisions to Evaluate customer continuity (see Table 4).

Table 4. Classification of customers according to the percentage of total sales

Class/Type	1	2	3
A	OC management and commercial executive approval	business meeting and debt letter	business meeting and notarial agreement
B	OC management and commercial executive approval	business meeting and deactivation letter	deactivation
C	OC management and commercial executive approval	business meeting and deactivation letter	deactivation

[a.] Type 1: Customers without provision. Type 2: Customers with a one-month supply. Type 3: Customers with provision (02) months

With this categorization, the information was organized to act in an orderly manner and with guide deadlines for their execution. The main objective is timely billing, favoring the corporation, especially in its cash flow.

With the 5S tool (see Fig. 4), in accordance with the proposed structure, an improvement has been made in base supervision. Although the context of services is different from that of an industrial process, the same environments are maintained with interaction in a space or field of work.

4.3 Application of the Case Study Model

Supervision is one of the most important operating procedures, as it allows incidents to be recognized and alerted to supervisors, managers, and customers. Currently, supervisors, on average, visit thirty agents per day according to the visiting guide assigned by the head of operations. With the updating of the administrative tasks of the supervisors

Fig. 4. Design 5S

(that is, requests for uniforms, equipment, implements, updating of courses, etc.), it is projected, for the first month, a liberation of 5% of their time, which will allow lengthen each supervision in approximately a minute and a half. Unexpected visits to jobs gave 100% implementation of the new model and 87% degree of satisfaction. With the application to the billing area, a table was managed where compliance with the tasks of the multidisciplinary team is monitored, with about twenty findings, highlighting the importance of deliverables and that 80% of inconsistencies are from A clients.

5   Discussion

5.1   New Scenarios vs Results

The execution time of the pilot was approximately 4 months, where three of them were executed as a pilot and one as a field work (see Table 5).

Table 5. Comparative: current situation vs. Final status

Indicators	Current situation	Final status
NPS Index 2022 vs 2023	36%	45%
Operational Detractors Index	75%	29%
Billing Detractors Index	43%	7%
Billing Error Rate	22%	6%
5S audit	22%	94%
supervision time	13%	19%

The main objective of this implementation was aimed at increasing the net promoter score, as such the project started with a 36% index, that is, the net between the percentages of promoters and detractors. After 4 months of implementation, it was possible to increase by 9% thanks to the reduction of detractors. Although it is a better indicator, it is important to mention that the number of customers classified as passive has been maintained. It is important to pay attention to it, they can be covert detractors (see Fig. 5).

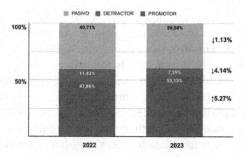

Fig. 5. Comparison of NPS 2022 vs 2023

On the other hand, it is important to conduct an analysis on the evaluation of the areas of operations and billing. Thanks to the second level that the satisfaction survey has, a comparison can be made and see how the implementation of the pilot projects has impacted.

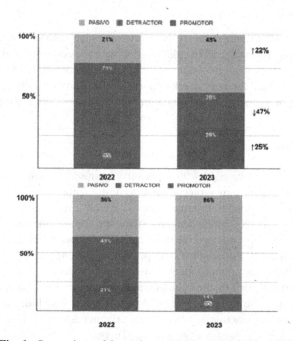

Fig. 6. Comparison of Operative and Billing NPS 2022 vs 2023

At the operations level, a considerable reduction of 46% of customers who were detractors and now no longer appear as such can be identified. This is thanks to the 25% increase in promoters, as well as a 22% increase in liabilities. At the KPI NPS level, it went from −71% to 1%. Although it is not an acceptable score in general aspects, it is a great advance for the brief time of implementation. At the billing level, a reduction of 36% can be identified. Obtaining with this a more efficient management of the delivery, as well as the assigned documentation. However, efficiency has neglected customers who were already promoters and made them passive. As mentioned above, it is a range of risk to have liabilities. The NPS score went from -22% to -14%. This is related to the percentage of errors reduced from 22% to 6%, reducing 16% (see Fig. 6).

The importance of the 5S tool is also enhanced by the correct recording of progress information. As stated in previous lines, a progressive improvement is identified in the weeks, starting from a critical situation of 22% to a report of 94%, this percentage is acceptable, but the continuity of the methodology must continue to be monitored.

Finally, supervision time increased by 6%, giving him approximately 3 min to add to each daily supervision he performs. In a meeting with the operational team, it was validated that quality supervision is the one that extends the time and considers segments oriented to the well-being of the collaborator. Important appreciation and finding since this can impact on a better-served human team, a supervisor with a better understanding of the field and therefore timely alerts, and a company with a positive impact on turnover, which in summary is cost reduction.

5.2 Future Works

Although the current management of the pilot was only 4 months, it is advisable to maintain the project for one year. In collaboration with the company, and in view of the results obtained, a work team has been created to provide continuity to the project. It is also important to extend the field of application. For this project, only the Lima region was used, 55% representation of the total number of collaborators at the national level.

6 Conclusions

Project management has brought with it various closing ideas that, if analyzed in the future, can be anchors for new projects. First, to destigmatize that the 5S and SCRUM tools should only be applied to product projects. We have reflected in this writing that the tools have been adapted to the different situations applied, in this case of services, it has helped to efficiently manage the daily actions of supervisors and billers in favor of improving their KPIs. Although it is important to use engineering tools applied to problems, it is vitally important to identify the most appropriate tool to take advantage of the information provided by the company as problematic. And that is translated into spending reduction. Here is the second point. Thirdly, it is concluded that there is not enough information or prior training on asset security management. That is, there are no specializations or careers that form professions. Therefore, it is challenging to stimulate the curriculum to important notions of service management.

Finally, it is concluded that service companies have a main asset linked to the person. For this reason, it is necessary to delve into human management and expose the impact of its mismanagement, since the fact of not controlling it can vary the profitability of the company.

References

1. Amati, M., et al.: Reducing changeover time between surgeries through lean thinking: an action research project. Front. Med. **9** (2022). https://doi.org/10.3389/fmed.2022.822964
2. Carrillo-Corzo, A., Tarazona-Gonzales, E., Quiroz-Flores, J., Viacava-Campos, G.: Lean process optimization model for improving processing times and increasing service levels using a deming approach in a fishing net textile company. In: Proceedings of the 6th Brazilian Technology Symposium (BTSym 2020), pp. 443–451 (2021). https://doi.org/10.1007/978-3-030-75680-2_50
3. Guimarães, R., et al.: Restructuring picking and restocking processes on a hypermarket. Prod. Eng. Arch. **28**(1), 64–72 (2022). https://doi.org/10.30657/pea.2022.28.08
4. Alavi, S., Aghakhani, H.: Identifying the effect of green human resource management practices on lean-agile (LEAGILE) and prioritizing its practices. Int. J. Prod. Perform. Manage. (2021). https://doi.org/10.1108/ijppm-05-2020-0232
5. Vieira, M.D., Azevedo, S.G., Pimentel, C.O., Matias, J.C.O.: Implementation of lean management system in a plastic packaging industry. Association for Computing Machinery (2022)
6. Boyle, S., Tyrrell, O., Quigley, A., Wall, C.: Improving ward level efficiency using a modified treatment room layout according to lean methodologies. Ir. J. Med. Sci. **191**(1), 127–132 (2021). https://doi.org/10.1007/s11845-021-02590-7
7. Calluche, F., Castillo, K., Viacava, G.: Improving construction waste management using transportation management and lean tools at a construction and hydrocarbons company. Institute of Electrical and Electronics Engineers Inc. (2022)
8. Synnes, E.L., Welo, T.: Using lean to transform the product development process in a marine company: a case study. Procedia CIRP **109**, 623–628 (2022). https://doi.org/10.1016/j.procir.2022.05.304
9. Diebold, P., Theobald, S., Wahl, J., Rausch, Y.: Stepwise transition to agile: from three agile practices to Kanban adaptation. J. Softw. Evol. Process **31**(5), e2167 (2019). https://doi.org/10.1002/smr.2167
10. Ruiz, L; Benítez, L; Pizo, A; Aristizabal, D.: Proposal for the optimization of the manufacturing process of integral kitchens of the Casa Madeira company across to lean six sigma methodology. IEOM Society (2021)
11. Arévalo, M., Montenegro, J., Viacava, G., Raymundo, C., Dominguez, F.: Proposal for process StandarDization for continuous improvement in a peruvian textile sector company. In: Ahram, T., Karwowski, W., Pickl, S., Taiar, R. (eds.) IHSED 2019. AISC, vol. 1026, pp. 909–915. Springer, Cham (2020). https://doi.org/10.1007/978-3-030-27928-8_136
12. Kadenic, M.D., Koumaditis, K., Junker-Jensen, L.: Mastering scrum with a focus on team maturity and key components of scrum. Inform. Softw. Technol. **153**, 107079 (2023). https://doi.org/10.1016/j.infsof.2022.107079
13. Ferigo, S., Tiraboschi, F., Rossi, M., Consagra, M.: Standardized work framework applied to ETO context. In: Rossi, M., Rossini, M., Terzi, S. (eds.) ELEC 2019. LNNS, vol. 122, pp. 407–416. Springer, Cham (2020). https://doi.org/10.1007/978-3-030-41429-0_40

14. Zarzycka, E., Dobroszek, J., Lepistö, L., Moilanen, S.: Coexistence of innovation and standardization: evidence from the lean environment of business process outsourcing. J. Manage. Control **30**(3), 251–286 (2019). https://doi.org/10.1007/s00187-019-00284-x

15. Junior, R.G.P., Inácio, R.H., da Silva, I.B., Hassui, A., Barbosa, G.F.: A novel framework for single-minute exchange of die (SMED) assisted by lean tools. Int. J. Adv. Manufact. Technol. **119**(9–10), 6469–6487 (2022). https://doi.org/10.1007/s00170-021-08534-w

Evaluation of Circular Business Models: Using a Multi Criteria Decision Analysis for Decision Support for Manufacturing SMEs

Simon Harreither[1,2(✉)], Fabian Holly[1], Constantin Magos[2],
and Günther Kolar-Schandlbauer[2]

[1] Institute of Management Science, TU Wien, Vienna, Austria
sharreither@efs.at
[2] EFS Consulting GesmbH, Vienna, Austria

Abstract. Companies are currently faced with the question which circular business models (CBMs) are best suited for their organization to achieve the highest possible circularity of their resources. Considering the complexity of the decision problem, small and medium-sized enterprises (SMEs) need a thorough approach to conquer this challenge. Since there is currently a research gap the aim of this study is to develop decision support for SMEs in the manufacturing sector in industrialized countries. The objective is to enable SMEs in choosing the most suitable CBM for their organization, while prioritizing their resources and activities by focusing on critical success factors (SFs) for the implementation of the CBMs. Based on a literature review, the five most frequently mentioned CBMs and twelve general SFs have been identified and evaluated using a multi criteria decision analysis method (MCDA) called Fuzzy PROMETHEE. The results display the rankings of the SFs and the CBMs. Deriving from the SFs ranking, it can be concluded that the most important factors for the successful development and implementation of CBMs are primarily aspects such as training, education and motivation, followed by cooperation and collaboration with all stakeholders and the separate allocation of internal financial resources. According to the ranking of the CBMs, it is recommended that SMEs should implement the Circular Supplies Model first followed by the Resource Recovery Model, Product Lifecycle Extension Model, Product-Service Systems and Sharing Platforms.

Keywords: circular economy · circular business models · success factors for circular business models · multi criteria decision analysis

1 Introduction

To meet the ecological, political and social challenges of climate change, an ideological rethinking of production and consumption is needed [1]. This requires alternatives to an economic system that relies on the continuous exploitation of primary resources. The circular economy (CE) represents such an alternative.[2] To successfully implement CE strategies in a company, an organizational transformation of existing business models

S.-H. Sheu (Ed.): IEIM 2024, CCIS 2070, pp. 183–203, 2024.
https://doi.org/10.1007/978-3-031-56373-7_15

(BM) to CBMs is necessary. These are based on minimizing resource utilization while generating as much economic value as possible.[3] In order to achieve the highest possible circularity of their resources, companies are currently faced with the question of which CBMs are best suited for their organization [4]. A number of decision support tools have already been developed in the relevant literature to facilitate the transition towards a CE. Among them are methods for checking the sustainability level of companies, tools for solving specific problems and scenario analysis, as well as studies comparing CE-related with linear behaviors.[5].

However, these studies only provide generalized statements on the selection of CBMs and make little to no reference to SMEs, whose BMs as well as entrepreneurial decisions are influenced by a variety of specific influencing factors and requirements [6]. Toker and Görener attempted to fill this research gap with their study. Based on a literature review and expert interviews, criteria for selecting four defined CBMs were identified and using the fuzzy TOPSIS ("Technique for Order Preference by Similarity to Ideal Solution") method, a recommendation for selecting CBMs for SMEs in developing countries was made. In their conclusion no optimum for all types of economy can be defined due to the continuous and increasingly intensive development of a CE. To prove or disprove their hypotheses, a study for industrialized countries is required.[4] The main objective of this work is to support manufacturing SMEs in industrialized countries to choose an appropriate CBM for their organization. The focus is on the optimal selection of CBMs and essential SFs that are necessary to consider. For this purpose, it is necessary to identify and evaluate relevant CBMs and general SFs for their implementation in manufacturing companies. Drawing upon the SFs, a recommendation will be formulated which CBMs are most suitable for the initial transition to a CE for manufacturing SMEs. Additionally, a recommendation will be made on how SMEs need to prioritize their resources and activities to successfully introduce CBMs.

2 Methodology

The methodology in this research is divided in four steps, which include a (1) literature review to present the state of research regarding CBMs (category 1) and their SFs (category 2), (2) data collection through problem centered interviews according to Witzel [7] to weight the SFs and to rate the CBMs against the SFs, (3) the application of the Fuzzy PROMETHEE („Preference Ranking Organization Method for the Enrichment of Evaluations") method according to Aytac et al. [8] and (4) the validation of the results.

2.1 Literature Review

In the course of the literature review, search strings (Table 1) were selected in combination with restriction of the observation period from 2018.

The sources listed in the respective literature databases were analyzed based on their titles and summaries. Publications that could not be excluded using the titles and abstracts were checked for their relevance to the objectives of this study based on their conclusion and full texts with which a basic literature list was compiled. Subsequently, the Backward and Forward method, according to Webster and Watson [9], was used

Table 1. Search strings of the literature review

Category 1	("Circular Economy" OR "Circularity") AND ("Circular Business Model" OR "Business Model") OR ("Circular Business Model Innovation")
Category 2	(("Circular Economy" OR "Sustainability") OR (("Circular Economy" AND "Business Models") OR "Circular Business Models") AND ("Drivers" OR "Barriers" OR "Enabling Factors" OR "Success Factors" OR "Critical Success Factors"))

on the basic literature list to search for further topic-specific studies. Table 2 lists the publications that were analyzed in this study. The total number of analyzed publications results from the basic literature list and the additional sources identified.

Table 2. Literature filtering process

Database / Search engine	No. of results	No. of results after analysis of titles and summaries	No. of studies in basic literature list	No. of found literature by the backward & forward method	No. Analyzed literature
Category 1					
ScienceDirect	4.071	11	6	11	32
Google Scholar	18.300	20	15		
Category 2					
ScienceDirect	4.817	10	8	21	42
Google Scholar	17.400	15	13		

As a primary scientific database, ScienceDirect was used. To identify a broader range of relevant sources, the search engine Google Scholar was involved, which, according to Halevi et al. [10], is suitable as an additional source for scientific literature due to its significant coverage expansions in recent years.

2.2 Data Collection

The required data to execute the MCDA were collected in two phases. In phase one, qualitative expert interviews were conducted with 15 CE experts who validated the SFs and weighted them based on the linguistic variables shown in Table 3. The number of interviewees is guided by Boddy's [11] study, which states that theoretical data saturation in qualitative research occurs with a sample size of at least twelve participants in a relatively homogeneous population. In phase two, also through qualitative interviews, the respective CBMs were evaluated against the SFs by five decision makers (DM) of SMEs in the manufacturing sector based on the linguistic variables, resulting in the ranking of the CBMs. There are generally no recommendations for the size of the expert panel in this phase of an MCDA, as no theoretical or empirical argumentation exists to prove a correlation between the sample size and more expressive results. Rather, the expertise of the participating individuals is important for the quality of the results.[12] By sending an informative sheet about the CBMs and the identified SFs, the interview participants were informed about the theoretical background of the work. At the beginning of the interviews, the initial situation, objectives and methodological approach of the evaluation method were presented. Selection problems require specific knowledge to ensure that priorities are set appropriately [13]. Those who set the criteria weights should be experts in their field to ensure that the weights are knowledgeable and well founded.

If the same individuals are responsible for both assigning weights to criteria and assessing alternatives, there is a potential for bias. By separating these tasks, more independent and objective results can be achieved.[14] In addition to this aspect, SME experts were used to evaluate the CBMs to bring their experience and expertise based on strategic decisions from an industrial perspective into this study. The profiles of the CE experts and DMs can be found in Table 12 and Table 13 in the appendix.

Table 3. Linguistic terms and corresponding trapezoidal fuzzy numbers.

Linguistic terms	Fuzzy numbers for weightings of SFs (a_1, a_2, a_3, a_4)	Fuzzy numbers for ratings of CBMs (a_1, a_2, a_3, a_4)
No importance (NI)	(0.0, 0.0, 0.1, 0.2)	(0.0, 0.0, 1.0, 2.0)
Low importance (LI)	(0.1, 0.2, 0.3, 0.4)	(1.0, 2.0, 3.0, 4.0)
Slight importance (SI)	(0.3, 0.4, 0.6, 0.7)	(3.0, 4.0, 6.0, 7.0)
Important (IM)	(0.6, 0.7, 0.8, 0.9)	(6.0, 7.0, 8.0, 9.0)
High importance (HI)	(0.8, 0.9, 1.0, 1.0)	(8.0, 9.0, 10, 10)

2.3 Fuzzy PROMETHEE

The PROMETHEE method is an outranking method and was developed by Brans in the early 1980s, in criticism of classical MCDA methods [15]. In the course of the PROMETHEE method, a finite number of discrete alternatives can be analyzed and evaluated based on pairwise comparisons. In this way, the intensity of the preferences of DMs can be represented. Compared to classical methods, PROMETHEE does not assume that, especially in uncertain decision environments, DMs are clearly aware of their preferences and are able to formulate them explicitly. To illustrate these uncertainties, preference functions are used to compare the alternatives with each other. This allows the expression of weak preferences or even incomparabilities with respect to the criteria.[16] These preference functions vary depending on the decision problem [17].

Another advantage of PROMETHEE is that by considering the internal relationships of each evaluation criterion during the decision process, complete compensation between the criteria is avoided [18]. This can be advantageous especially for decision problems concerning questions of ecology and thus also sustainability and CE [16]. The main challenge of MCDAs is to transform the qualitative observations and preferences made by individuals into quantitative input data [19]. The complexity of MCDA problems arises from the decision environment. This is due to the lack of specific knowledge about the objectives and constraints and the difficulty to describe the problem explicitly in precise values.[20] To assist in this problem and to obtain the desired results from uncertain and ambiguous data, the use of fuzzy set theory developed by Zadeh [21] is an appropriate measure. This approach links linguistic variables with numerical fuzzy sets to represent linguistic expressions mathematically [21]. The fundamental elements of fuzzy set theory consist of the membership functions, which visualize the fuzzy sets, and the aggregation operators, that are used for the mathematical combination of fuzzy sets [22].

The form of the membership function depends on the type of fuzzy numbers. In the relevant literature, there are several different approaches, including triangular, trapezoidal and gaussian fuzzy numbers.[23] While gaussian fuzzy numbers are suitable for problems dealing with probabilities and statistical data, triangular and trapezoidal fuzzy numbers are appropriate for the representation of variables that are based on imprecise or qualitative data [24]. Although triangular fuzzy numbers are a simple method to deal with insufficient or inaccurate data, this concept faces the problem that due to the triangular shape of the membership function, only one point can be used to represent the most probable value. In practice, however, it is often not possible to define a unique peak value, which means that the application of triangular fuzzy numbers often leads to significant simulation errors that can affect the results. A trapezoidal membership function, on the other hand, has similar advantages to a triangular and is characterized by its adaptability for parameter distributions over intervals of peaks. That means trapezoidal fuzzy numbers take into account at least two peaks, which can easily eliminate the problem of simulation errors.[25] For this reason, trapezoidal fuzzy numbers were used in this paper.

2.4 Validation

The validation involves a local sensitivity analysis and a comparative analysis to check the robustness and reliability of the results. The objective of the sensitivity analysis is to investigate the influences of the criteria weights on the evaluation of the CBMs [4]. The comparative analysis is performed by using two alternative MCDAs called Fuzzy ELECTRE I according to Hatami-Marbini & Tavana [26] and Fuzzy TOPSIS according to Chen et al. [27].

3 Results

3.1 Literature Review

Circular Business Models. The numerous challenges of an ecological and social nature facing the current economic system, such as climate change as well as the progressive destruction of the environment and the scarcity of natural resources are an expression of the lack of sustainability of the production and consumption models that have been common practice up to now [28]. To effectively address these difficulties, companies will have to drastically change their operations. Transforming their BMs towards CBMs can play a significant role within this change.[29] CBMs are defined as models in which value creation is based on the use of residual economic value to produce new products. They provide a conceptual framework for returning discarded products from the user to the producer, thus closing supply and material loops. CBMs include approaches to reduce the use of resources as well as waste and emissions. These approaches encompass cycling, extending, intensifying and dematerializing material and energy loops.[3] Currently, five CBMs that support the transition to a CE are discussed in the literature. Individually or in combination, the identified CBMs offer a viable approach to implement CE strategies and contribute to a successful transformation from a linear economy to a CE due to their potential to minimize or eliminate waste, emissions, and inefficiencies. Focused and intensive application is required to exploit the potential of CBMs with maximum effect. Nevertheless, the CBMs are not mutually exclusive and are characterized by their compatibility with each other.[30] These CBMs cover the entire value chain with each of the CBMs starting at different stages and in some cases building on each other. While (1) Circular Supplies Models (CSM) and (2) Resource Recovery Models (RRM) focus on the procurement and manufacturing processes, (4) Product-Service Systems (PSS) and (5) Sharing Platforms (SP) target the consumption and use phases of the products. The (3) Product Lifecycle Extension Model (PLE) represents a CBM that focuses on both the manufacturing and the use of the products.[30] The different CBMs and their characteristics can be seen in Table 4.

Table 4. The five Circular Business Models

CBM	Characteristics
Circular Supplies Model (CSM)	Substitution of critical resources by circular alternatives, such as renewable energies, renewable, biobased or recyclable materials [31]
Resource Recovery Model (RRM)	Recovery of resources or energy from waste or end-of-life products [31]
Product Lifecycle Extension Model (PLE)	Extension of product use through design measures and repair and maintenance services [32]

(continued)

Table 4. (*continued*)

CBM	Characteristics
Product-Service Systems (PSS)	Linking products with services. The manufacturer retains ownership of the products and offers them to customers for a specific period of use [33]
Sharing Platforms (SP)	Optimal utilization of resources by intersectional sharing or collaborative models for use, access or ownership [34]

Success Factors for the implementation of Circular Business Models. SFs are characteristics of a company or its environment that have a significant positive influence on the company's success [35]. This indicates that SFs are the measures and activities that contribute significantly to the future success of a company. Providing the SFs for CBMs enables companies to prioritize appropriate actions and thus reduce management effort, as well as increase the probability of success and management efficiency in the selection, development and implementation of CBMs.[36] As a result of the literature review, twelve general qualitative SFs were identified, which were categorized into five dimensions. These dimensions are (1) products and services, (2) technology and infrastructure, (3) resources, (4) network and partnerships, and (5) economic efficiency and financial sustainability. The SFs can be seen in Table 5.

Table 5. Success Factors for the implementation of Circular Business Models

Group	Success Factor		Brief description	Exemplary references
Products & Services	SF_1	Design of products for circularity	Design practices for (1) product durability & resilience, (2) standardization & compatibility, (3) recyclability, (4) ease of maintenance, repair, disassembly, and reassembly, (5) adaptability & upgradability, (6) binding and trust building	[30, 37–40]
	SF_2	Improved functionality & longer product life through services	Evaluation of service agreements for (1) maintenance, repair, restoration, (2) provision of information for cleaning, installation and basic maintenance processes for consumers	[38, 39, 41–43]

(*continued*)

Table 5. (*continued*)

Group	Success Factor		Brief description	Exemplary references
Technology & Infrastructure	SF$_3$	Technologies to mechanize circular economy strategies (9R)	Technological modernization, efficient technologies for R-strategies (Reuse, Repair, Remanufacturing, Refurbish, Recycling, et cetera). For instance, robotics, 3D printing, sensor technology for waste or material sorting, shredding systems, et cetera	[38, 44–47]
	SF$_4$	Digital technologies	Technologies of IoT, AI, Blockchain, et cetera for (1) monitoring of real-time data (tracking, digital product/ material passes), (2) automation, (3) predictive/ prescriptive maintenance (4) communication & information flow, (4) product & process planning	[30, 41, 44, 45, 47]
	SF$_5$	Separate waste collection & product return systems	Collection and separation of critical raw materials, return of end-of-life products	[41, 48–51]
Resources	SF$_6$	Use of renewable energies	Reducing emissions from fossil energy sources, ensuring the environmental benefits of CBMs	[40, 52–55]
	SF$_7$	Use of renewable & recyclable materials	Substitution of critical raw materials with renewable, bio-based materials, recyclable materials	[38, 40, 46, 56]
	SF$_8$	Optimized use of resources & waste reduction	Increase in material and energy efficiency through (1) quality improvements, (2) product and process optimization, (3) higher plant and machine utilization, (4) reduction of inventories	[38, 45, 57–59]

(continued)

Table 5. (*continued*)

Group	Success Factor		Brief description	Exemplary references
Network & Partnerships	SF$_9$	Training, education & motivation	CE-oriented education and motivation of suppliers, partners, customers & employees, continuous training & qualification of employees	[38, 44, 45, 57, 59, 60]
	SF$_{10}$	Cooperation & collaboration with all stakeholders	Establish networks & partnerships, continuous knowledge exchange, collaboration with politics to co-create policies and regulations	[38, 41, 44–46, 61]
Economic efficiency & financial sustainability	SF$_{11}$	Separate allocation of internal financial resources	Budgets for (upfront) investments for the transition to a circular production/business model. E.g., for R&D, infrastructure, new technologies, et cetera	[38, 44, 45, 62, 63]
	SF$_{12}$	Financial support from stakeholders and politics	Loans, tax incentives and financial incentives for recyclable products, subsidies, and refunds	[30, 44–46, 61]

3.2 Data Collection

Weightings of the Success Factors. In the further course, the SFs were validated and weighted by the experts on basis of their experiences and subjective perceptions. The weightings of the respective SFs are illustrated in Table 6.

Table 6. Weightings of the Success Factors

Ci	EX$_1$	EX$_2$	EX$_3$	EX$_4$	EX$_5$	EX$_6$	EX$_7$	EX$_8$	EX$_9$	EX$_{10}$	EX$_{11}$	EX$_{12}$	EX$_{13}$	EX$_{14}$	EX$_{15}$
SF$_1$	HI	HI	HI	HI	HI	HI	HI	HI	HI	LI	HI	HI	HI	HI	HI
SF$_2$	HI	HI	SI	HI	IM	HI	IM	HI	HI	IM	IM	IM	IM	SI	HI
SF$_3$	SI	IM	SI	SI	HI	IM	SI	IM	HI	SI	IM	IM	SI	IM	IM
SF$_4$	IM	IM	SI	IM	HI	HI	SI	IM	HI	IM	HI	SI	IM	IM	HI
SF$_5$	HI	IM	IM	LI	SI	IM	HI	IM	HI	SI	IM	HI	HI	HI	SI
SF$_6$	SI	HI	HI	IM	SI	SI	LI	HI	IM	LI	HI	IM	SI	IM	HI

(*continued*)

Table 6. (*continued*)

Ci	EX$_1$	EX$_2$	EX$_3$	EX$_4$	EX$_5$	EX$_6$	EX$_7$	EX$_8$	EX$_9$	EX$_{10}$	EX$_{11}$	EX$_{12}$	EX$_{13}$	EX$_{14}$	EX$_{15}$
SF$_7$	IM	IM	HI	IM	HI	HI	IM	SI	HI	LI	HI	HI	HI	HI	HI
SF$_8$	IM	IM	SI	SI	IM	HI	SI	HI	HI	IM	HI	LI	IM	HI	HI
SF$_9$	HI	HI	HI	HI	IM	HI	SI	HI	HI	IM	HI	HI	HI	IM	HI
SF$_{10}$	HI	HI	HI	IM	IM	IM	SI	IM	HI	HI	HI	HI	IM	HI	IM
SF$_{11}$	HI	IM	HI	SI	HI	SI	IM	IM	HI	HI	HI	HI	IM	HI	IM
SF$_{12}$	IM	IM	HI	SI	IM	SI	IM	IM	HI	IM	IM	HI	IM	IM	IM

Ratings of the Circular Business Models. The rating of the CBMs was carried out in the context of which SFs the DMs would rely on for the respective CBMs to implement them successfully. The aim here was to incorporate the practice-oriented expertise and experience regarding strategic decisions of the DMs of SMEs into the evaluation. Table 7 shows the ratings of the DMs.

Table 7. Ratings of the Circular Business Models by the decision makers of SMEs

DMi	CBMs	SF$_1$	SF$_2$	SF$_3$	SF$_4$	SF$_5$	SF$_6$	SF$_7$	SF$_8$	SF$_9$	SF$_{10}$	SF$_{11}$	SF$_{12}$
DM$_1$	**CSM**	HI	HI	HI	IM	HI	HI	HI	HI	IM	HI	HI	IM
	RRM	HI	SI	HI	IM	HI	HI	HI	IM	IM	SI	SI	SI
	PLE	HI	HI	SI	HI	IM	HI	IM	HI	HI	SI	SI	SI
	PSS	SI	HI	SI	HI	HI	HI	HI	HI	HI	SI	SI	SI
	SP	SI	IM	LI	HI	SI	HI	HI	HI	IM	IM	SI	SI
DM$_2$	**CSM**	IM	IM	SI	HI	LI	HI	HI	HI	SI	IM	HI	HI
	RRM	HI	LI	LI	HI	HI	IM	HI	LI	LI	SI	HI	HI
	PLE	HI	HI	LI	HI	NI	SI	LI	IM	HI	IM	HI	HI
	PSS	SI	HI	LI	HI	NI	IM	SI	LI	IM	HI	HI	HI
	SP	SI	IM	NI	HI	NI	IM	LI	HI	IM	HI	HI	HI
DM$_3$	**CSM**	HI	SI	IM	HI	IM	HI	HI	HI	IM	IM	SI	HI
	RRM	HI	HI	HI	IM	HI	SI	HI	IM	IM	LI	IM	SI
	PLE	HI	HI	IM	IM	LI	SI	IM	IM	IM	IM	IM	SI
	PSS	IM	HI	SI	HI	LI	SI	HI	HI	HI	HI	IM	LI
	SP	SI	IM	SI	HI	SI	SI	SI	SI	IM	HI	IM	IM
DM$_4$	**CSM**	IM	HI	HI	HI	HI	HI	HI	HI	HI	SI	HI	HI
	RRM	IM	IM	IM	HI	HI	IM	HI	HI	HI	IM	HI	HI

(*continued*)

Table 7. (*continued*)

DMi	CBMs	SF$_1$	SF$_2$	SF$_3$	SF$_4$	SF$_5$	SF$_6$	SF$_7$	SF$_8$	SF$_9$	SF$_{10}$	SF$_{11}$	SF$_{12}$
	PLE	HI	HI	HI	HI	SI	HI	IM	HI	HI	IM	IM	IM
	PSS	IM	HI	IM	IM	SI	IM	SI	IM	IM	IM	IM	IM
	SP	SI	HI	SI	HI	LI	LI	LI	IM	IM	HI	IM	IM
DM$_5$	CSM	HI	LI	SI	SI	HI	HI	HI	IM	IM	IM	IM	SI
	RRM	HI	NI	HI	IM	HI	LI	HI	LI	LI	SI	HI	IM
	PLE	HI	HI	IM	IM	NI	NI	SI	SI	HI	SI	LI	IM
	PSS	NI	IM	NI	HI	LI	NI	NI	IM	HI	LI	HI	LI
	SP	SI	IM	NI	HI	NI	NI	NI	HI	IM	IM	SI	LI

3.3 Application of Fuzzy PROMETHEE

Priority Ranking of the Success Factors. Using the respective fuzzy numbers (Table 3), the weightings of the CE experts (Table 6) were aggregated and defuzzified [8]. The normalized weightings were determined by the normalization technique of the linear sum. For each SF, the normalized weighting results from the ratio of the respective defuzzified weightings and the sum of all defuzzified weightings. The ranking of the SFs is derived by the size of the normalized weightings.[64] The results can be seen in Table 8. According to the results, SFs such as (SF$_9$) training, education & motivation, (SF$_{10}$) cooperation and collaboration with all stakeholders, and (SF$_{11}$) separate allocation of internal financial resources were rated best. Factors such as (SF$_5$) separate waste collection and product return systems, the (SF$_8$) optimized use of resources and waste reduction, and the (SF$_6$) use of renewable energies were rated lowest.

Table 8. Priority ranking of the Success Factors

Ci	Aggregated fuzzy weights	Defuzzified weights	Normalized weights	Rank
SF$_1$	(0.10, 0.85, 0.95, 1.00)	0.727	0.0867	5
SF$_2$	(0.30, 0.75, 0.87, 1.00)	0.730	0.0871	4
SF$_3$	(0.30, 0.61, 0.75, 1.00)	0.663	0.0792	9
SF$_4$	(0.30, 0.71, 0.83, 1.00)	0.708	0.0846	6
SF$_5$	(0.10, 0.69, 0.81, 1.00)	0.648	0.0774	10
SF$_6$	(0.10, 0.62, 0.75, 1.00)	0.617	0.0736	12
SF$_7$	(0.10, 0.77, 0.87, 1.00)	0.685	0.0818	8
SF$_8$	(0.10, 0.69, 0.81, 1.00)	0.648	0.0774	11

(*continued*)

Table 8. (*continued*)

Ci	Aggregated fuzzy weights	Defuzzified weights	Normalized weights	Rank
SF_9	(0.30, 0.83, 0.93, 1.00)	0.765	0.0913	1
SF_{10}	(0.30, 0.79, 0.89, 1.00)	0.745	0.0889	2
SF_{11}	(0.30, 0.77, 0.88, 1.00)	0.737	0.0879	3
SF_{12}	(0.30, 0.70, 0.81, 1.00)	0.703	0.0840	7

Ranking of the Circular Business Models. According to the procedure following Aytac et al. [8], based on the fuzzy preference index per alternative, the defuzzified leaving (Φ^+) and entering flows (Φ^-) as well as net flows (Φ^{NET}) per alternative are obtained as listed in Table 9. While the leaving flow expresses the degree of dominance of an CBM over the other CBMs regarding the SFs, the entering flow represents the extent to which an alternative is dominated by the respective other alternatives. The value of the respective net flows provides information on the outranking relationship and the ranking order of the CBMs.[16] The results demonstrate that the CSM, is the best rated CBM, followed by RRM, PLE, PSS, and SP. This conclusion is based on the net flows which are determined by the differences between the leaving and entering flows of each CBM [8]. The CSM is the CBM with the largest leaving flow and the smallest entering flow. This means that the CSM dominates the other CBMs in terms of most SFs. In contrast, SPs are outweighed by all other CBMs due to the smallest net current.

Table 9. Defuzzified flows and ranking of the Circular Business Models

Φ	Φ^+	Φ^-	Φ^{NET}	Rank
CSM	2.760	0.022	2.738	1
RRM	2.632	0.432	2.200	2
PLE	1.054	1.403	-0.350	3
PSS	0.458	1.994	-1.537	4
SP	0.051	3.103	-3.052	5

3.4 Validation

Sensitivity Analysis. In the process of the sensitivity analysis, eight scenarios were created, each with different values of the aggregated fuzzy weights of the criteria. The first scenario (S_1) represents a ten percent reduction in the fuzzy numbers of the six SFs with the highest weights (SF_9, SF_{10}, SF_{11}, SF_2, SF_1, SF_4). In the second scenario (S_2), the six weakest SFs (SF_{12}, SF_7, SF_3, SF_5, SF_8, SF_6) are increased by ten percent, and the third scenario (S_3) considers the changed fuzzy weights of both S_1 and S_2. Scenarios

S_4, S_5, and S_6 are set analogous to scenarios S_1, S_2, and S_3 with a 20 percent reduction and increase in values, respectively. While the seventh scenario (S_7) considers only the six SFs with the highest weights in the initial state (S_0) in the analysis, the focus of the eighth scenario (S_8) is on the original values of the six weakest SFs. The scenarios considered are shown in Table 10.

Table 10. Tested scenarios of the sensitivity analysis

SF	S_1	S_2	S_3	S_4	S_5	S_6	S_7	S_8
SF_9, SF_{10}, SF_{11}, SF_2, SF_1, SF_4,	90%	100%	90%	80%	100%	80%	100%	0%
SF_{12}, SF_7, SF_3, SF_5, SF_8, SF_6,	100%	110%	110%	100%	120%	120%	0%	100%

As shown in Fig. 1 the net flows of all CBMs remain at a constant level in six of the eight scenarios. The ranking of CBMs remains the same in all eight scenarios. While the net flows of alternatives PLE, PSS, and SP increase in S_7, a reduction to approximately the same value results for CSM and RRM. It can be concluded that the criteria SF_{12}, SF_7, SF_3, SF_5, SF_8 and SF_6 have a positive influence on the assessment of the alternatives CSM and RRM and negative effects on PLE, PSS and SP. For S8, an opposite trend of the net flows is visible. While neglecting the strongest SFs leads to a strong increase in net flows for CSM and RRM and a strong decrease in net flow for SP, the scores of PLE and PSS remain at a similar level as in the first six scenarios. This indicates that criteria SF_9, SF_{10}, SF_{11}, SF_2, SF_1, and SF_4 negatively affect the scores of CSM and PSS, and positively affect the scores of PLE, PSS, and SP.

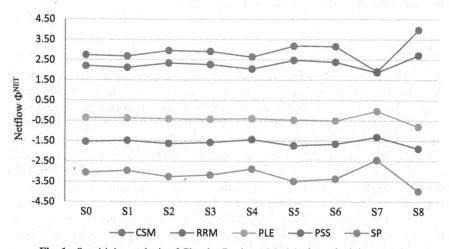

Fig. 1. Sensitivity analysis of Circular Business Models through eight scenarios

Comparative Analysis. In this section, the comparative analysis carried out to check the reliability of the results is illustrated [4]. For this purpose, two alternative MCDA methods, fuzzy ELECTRE I and fuzzy TOPSIS, were additionally used for the evaluation of the CBMs. The results can be seen in Table 11. Clearly, CSM is the best rated CBM among all methods.

Table 11. Comparison of the rankings of the MCDA methods used.

	Fuzzy PROMETHEE	Fuzzy ELECTRE I	Fuzzy TOPSIS
CSM	1	1	1
RRM	2	3	2
PLE	3	2	3
PSS	4	4	4
SP	5	4	5

The comparative analysis shows that similar results are obtained from all three MCDA methods used. This indicates that the results of the proposed evaluation method are reliable. Only fuzzy ELECTRE I shows partially different outcomes. This could be since when determining the Boolean concordance and discordance matrices, which form the basis for the pairwise comparison of the alternatives in ELECTRE I, the exact values are transformed into Boolean numbers. As a result, the degree of concordance or discordance is only differentiated between zero and one or "weak" and "strong". Fuzzy PROMETHEE and Fuzzy TOPSIS use continuously exact values, which subsequently could lead to clearer differences between the alternatives.

4   Discussion

To ensure the success of the initiative, it is recommended by the authors of this study that SMEs should first implement a CBM that is best suited to their business characteristics. Once this CBM is viable, another CBM can be developed and implemented building on it. However, SMEs are subject to the problem of not being able to invest in several innovation projects at the same time due to their limited resources and financial means. This also affects the implementation of CBMs.[65] The greatest possible effect in implementing a CBM is theoretically achieved by focusing on all identified SFs simultaneously. However, in practice, simultaneous engagement in multiple SFs or the wrong order of priorities, especially for SMEs, can lead to difficulties and, in the worst case, to counteracting effects, which could reduce or even prevent the success of the initiative. In addition, a simultaneous introduction has the consequence that a concurrent, enormous change process takes place in the company as well as extensive resources are required for the change. Ranking the SFs can assist with this challenge and provides an indication of the order in which the SFs should be prioritized to ensure the greatest success in implementing CBMs, strengthen acceptance for change, and make the best use of necessary resources during the transition.

The results of this study demonstrate that for the successful implementation of CBMs, priority must be given to aspects such as (SF_9) training, education and motivation, (SF_{10}) cooperation and collaboration, and (SF_{11}) separate allocation of internal financial resources in the first stages of the transformation process. The CE-oriented education and training of employees, the education and motivation of suppliers and partners as well as the intensive cooperation with all internal and external stakeholders lay the foundation for the success of CBMs. Especially SMEs are often not capable of building up circular competencies and capabilities or developing circular processes or closed loop supply chains on their own. For this reason, it is necessary to make employees fit for the CE and to cooperate intensively with all upstream and downstream stakeholders of the value chain to develop these structures. However, according to the experts, all SFs discussed have a non-negligible relevance for CBMs.

Based on ranking of the CBMs, the authors recommend that SMEs in the manufacturing sector should focus on CSM and RRM, followed by the PLE, PSS and SP for the transition to a CE. These results are similar to the study by Toker and Görener, except that RRM leads the ranking. They emphasize for SMEs in developing countries the first steps are to increase resource and waste efficiency and to create a recycling society.[4] Owolabi et al. [66] have shown in their study that companies in developing countries have less mature solid waste management systems than industrialized countries due to insufficient and ineffective policies. This may explain the difference between the rankings for SMEs in developing and industrialized countries.

However, the ranking does not mean that one CBM excludes the others. The focus of CSM is on avoiding waste and substituting critical resources with circular alternatives such as renewable resources or recyclable materials [30]. Based on the characteristics of the other CBMs, the CSM can serve as the foundation for introducing the other CBMs in the order of the ranking. Accordingly, the CSM lays the base for circular products, enabling the reuse or recycling of their materials. Once the CSM has been successfully implemented, for products that have reached their end of life and whose materials are no longer suitable for reuse the possibility arises to introduce the RRM. This involves the creation of secondary raw materials or recycled materials from waste streams and end-of-life products. After closing the resource loops by the CSM and RRM, the focus can be placed on CBMs, which address the use phase and have the potential to slow down the resource loops. The starting point for this marks the PLE. This model takes on measures in the design and use phases to extend product life, as well as to maintain, repair and upgrade products. A longer product life leads to a reduced need for new products and thus to a reduction in the dependence of economic growth on the use of resources. This can be particularly challenging for sales-oriented SMEs that depend on short product cycles, as longer-lived products often result in lower sales volumes. The introduction of additional service agreements for the maintenance, repair or upgrading of products could compensate for any resulting short- to midterm revenue losses and build a bridge towards service-oriented CBMs like PSS. PSS involves providing products as services or with services, shifting ownership to manufacturers. This represents a paradigm shift from sales to service orientation, necessitating strategic reorientation for companies and changes in consumption patterns.

According to the study by Thompson et al. [67], the PLE could lay the foundation for the introduction of PSS. If products that have been developed for the longest possible lifespan are supplemented by a PSS, the opportunity arises for the manufacturer to gain a competitive advantage [67]. That means the manufacturer, which provides its products as a service, has an interest in making the products durable and of high quality so that it can derive the greatest possible value from them in the long term. An advantage of PSS is that they can be built in parallel to an existing BM to facilitate the transformation. Finally, after closing the resource loops through CSM and RRM and after the shift to service orientation by PLE and PSS, the opportunity arises for SMEs to intensify the use of existing resources by intersectional sharing through SPs. In addition, SPs offer the option of making PSS and PLE products and services available to a larger group of users in a simple way.

When performing the comparative analysis with Fuzzy ELECTRE and Fuzzy TOP-SIS, similar results were obtained. The sensitivity analysis showed that the ranking of CBMs remains the same in all tested scenarios. This proves the robustness and reliability of the results presented in this paper.

5 Conclusion

This study provides an overview of CBMs and critical SFs for their successful introduction. Providing the SFs can help academics and SMEs to generate a deeper understanding of CBMs and assess their opportunities, risks, challenges, and feasibility. On this basis, management can formulate an effective strategy for the systematic introduction of a specific CBM. Furthermore, the results of this study provide theoretical indications for the scientific discussion on the CE.

The results of this study have led to the ranking of SFs and CBMs, from which a recommendation on their prioritization could be made. By interviewing CE experts and DMs in the context of the needs of SMEs in the manufacturing sector, the specialization of the MCDA results for manufacturing SMEs emerges. Nevertheless, the SFs have a general character and can serve as orientation for other company sizes and industries. However, it should be noted that, for example, in the context of large companies or other industries, other prioritizations of the SFs or CBMs could result.

It should be noted that these rankings depend on the subjective judgments of the interview participants. This means that the outcome of the rankings depends on the situation and on specific considerations of the participants that motivate the adoption of CBMs. For example, if the assessments are primarily conducted from the perspectives of corporate social responsibility, environmental priorities or sustainability aspects, a different ranking could result. Furthermore, the sample size of five interviewed SMEs in the manufacturing sector represents a limitation. Although no theoretical or empirical arguments exist for conducting MCDAs that demonstrate a correlation between sample size and the validity of the results [12], theoretical data saturation in qualitative research occurs with a minimum number of twelve participants [11]. Accordingly, the ranking of CBMs cannot be generalized and only a recommendation can be made on which CBMs SMEs in the manufacturing sector should focus for the first steps towards a CE. Future studies could follow up on this paper by examining the correlations of SFs and CBMs in a broader sample size to test the generalizability of the results. To deepen the research on the selection of CBMs for SMEs, the identification of external criteria that influence SMEs in the selection of CBMs would also be relevant for further studies. This can help to shed light on the decision-making environment of SMEs and generate a coherent understanding of external factors influencing CBMs.

Appendix

Table 12. Profiles of the decision makers of the SMEs

DMi	Position	Industry	Number of employees
DM_1	Managing Director	Automotive	~ 350
DM_2	Managing Director	Plant Engineering	< 50
DM_3	Representative for EQS[1]	Automotive	~ 170
DM_4	Managing Director	Mechanical and Plant Engineering	< 200
DM_5	Managing Director	Mechanical Engineering	< 50

Table 13. Profiles of the Circular Economy Experts

EXi	Position
EX_1	Research in the field of assembly planning and assistance systems in the research area of work design and digitalization with research interests in CE
EX_2	Research in the field of social ecology at an Austrian university
EX_3	CE expert, Executive Director and Academic Program Director for sustainable finance at a CE Forum
EX_4	CE expert and consultant
EX_5	Head of the research department for waste management and resource management at an Austrian university
EX_6	Representative of the Federal Ministry for Climate Protection, Environment, Energy, Mobility, Innovation and Technology of Austria
EX_7	Director CE of an international company in the fields of auditing, tax consultancy, legal advice and business or management consultancy based in Austria
EX_8	University lecturer at the Institute for Manufacturing and Photonic Technologies of an Austrian university with research interests in CE

(*continued*)

---

[1] Environment, Quality and Occupational Safety on behalf of the Managing Director.

Table 13. (*continued*)

EXi	Position
EX9	Global Lead CE and Packaging of a German company in the consumer goods and adhesives industry
EX10	Head of department in the CE division of a German non-profit research institution focusing on climate, environmental and energy issues
EX11	Co-founder, Managing Director and Key Researcher of an Austrian company specializing in market research and consulting with research interests in CE
EX12	Professor for Business Model Design at an Austrian University and Advisor for Sustainable Business Models
EX13	Research associate at an Austrian university with research interests in CE and social network analysis
EX14	Research assistant at an Austrian university with research interests in operations engineering, systems planning and facility management, human-machine interaction and CE
EX15	Product expert for CE at an Austrian certification organization

References

1. Europäische Kommission. Ein neuer Aktionsplan für die Kreislaufwirtschaft für ein saubereres und wettbewerbsfähigeres Europa. https://eur-lex.europa.eu/legal-content/DE/TXT/?qid=1583933814386&uri=COM%3A2020%3A98%3AFIN. Accessed 02 Apr 2022
2. Herrmann, C., Vetter, O.: Ökologische und ökonomische Bewertung des Ressourcenaufwands: Remanufacturing von Produkten. VDI Zentrum Ressourceneffizienz GmbH (2021)
3. Geissdoerfer, M., Pieroni, M.P.P., Pigosso, D.C.A., Soufani, K.: Circular business models: a review. J. Clean. Prod. **277**, 123741 (2020). https://doi.org/10.1016/j.jclepro.2020.123741
4. Toker, K., Görener, A.: Evaluation of circular economy business models for SMEs using spherical fuzzy TOPSIS: an application from a developing countries' perspective. Environ. Dev. Sustain. **25**, 1700–1741 (2022)
5. Rosa, P., Sassanelli, C., Terzi, S.: Towards circular business models: a systematic literature review on classification frameworks and archetypes. J. Clean. Prod. **236**, 117696 (2019). https://doi.org/10.1016/j.jclepro.2019.117696
6. Becker, W., Krämer, J., Ulrich, P.: Typologie mittelständischer Unternehmen. ZfO **82**, 348–352 (2013)
7. Witzel, A.: Das problemzentrierte interview. Forum Qualitative Sozialforschung, 1(1) (2000)
8. Aytac, E., Tuşlşık, A., Nilsen Karakasoglu, K.: An alternative approach based on Fuzzy PROMETHEE method for the supplier selection problem. Uncertain Supply Chain Manage. **4**(3), 183–194 (2016)
9. Webster, J., Watson, R.T.: Analyzing the past to prepare for the future: writing a literature review. MIS Q. **26**(2), 13–23 (2002)
10. Halevi, G., Moed, H., Bar-Ilan, J.: Suitability of Google Scholar as a source of scientific information and as a source of data for scientific evaluation - review of the literature. J. Informet. **11**(3), 823–834 (2017)

11. Boddy, C.R.: Sample size for qualitative research. Qual. Mark. Res. Int. J. **19**(4), 426–432 (2016). https://doi.org/10.1108/QMR-06-2016-0053
12. Bulut, E., Duru, O.: Analytic hierarchy process (AHP) in maritime logistics: theory, application and fuzzy set integration. In: Lee, P.T.-W., Yang, Z. (eds.) Multi-Criteria Decision Making in Maritime Studies and Logistics, pp. 31–78. Springer International Publishing, Cham (2018). https://doi.org/10.1007/978-3-319-62338-2_3
13. Ivlev, I., Kneppo, P., Barták, M.: Method for selecting expert groups and determining the importance of experts' judgments for the purpose of managerial decision-making tasks in health system. E+M Ekonomie a Manage. **18**(2), 57–72 (2015). https://doi.org/10.15240/tul/001/2015-2-005
14. Montibeller, G., von Winterfeldt, D.: Biases and debiasing in multi-criteria decision analysis. In: 48th Hawaii International Conference on System Sciences (2015)
15. Brans, J.-P., Bertrand, M.: Multiple Criteria Decision Analysis: State of the Art Surveys: Promethee Methods (2005)
16. Geldermann, J., Nils, L.: Leitfaden zur Anwendung von Methoden der multikriteriellen Entscheidungsunterstützung: Methode: PROMETHEE https://www.uni-goettingen.de/de/document/download/285813337d59201d34806cfc48dae518-en.pdf/MCDA-Leitfaden-PROMETHEE.pdf. Accessed 17 Mar 2023
17. Brans, J.P., Vincke, P., Mareschal, B.: How to select and how to rank projects: the Promethee method. Eur. J. Oper. Res. **24**(2), 228–238 (1986)
18. Murat, S., HalimKazan, S., Coskun, S.: An application for measuring performance quality of schools by using the promethee multi-criteria decision making method. Proc. Soc. Behav. Sci. **195**, 729–738 (2015). https://doi.org/10.1016/j.sbspro.2015.06.344
19. Goumas, M., Lygerou, V.: An extension of the PROMETHEE method for decision making in fuzzy environment: ranking of alternative energy exploitation projects. Eur. J. Oper. Res. **123**(3), 606–613 (2000)
20. Bellman, R.E., Zadeh, L.A.: Decision-making in a fuzzy environment. Manage. Sci. **17**(4), 141–164 (1970)
21. Zadeh, L.A.: Fuzzy sets. Inf. Control **8**(3), 338–353 (1965)
22. Poincaré, H.: Fuzzy-set-theorie. In: Keuper, F. (ed.) Fuzzy-PPS-Systeme, pp. 63–118. Deutscher Universitätsverlag, Wiesbaden (1999). https://doi.org/10.1007/978-3-322-993 15-1_3
23. Chakraverty, S., Sahoo, D.M., Mahato, N.R.: Fuzzy numbers. In: KChakraverty, S., Sahoo, D.M., Mahato, N.R. (eds.) Concepts of Soft Computing, pp. 53–69. Springer, Singapore (2019). https://doi.org/10.1007/978-981-13-7430-2_3
24. Adil, O., Ali, A.: Comparison between the effects of different types of membership functions on fuzzy logic controller performance. Int. J. Emerg. Eng. Res. Technol. **3**(4), 76–83 (2015)
25. Liu, G., Wang, X.: A trapezoidal fuzzy number-based vikor method with completely unknown weight information. Symmetry **15**(2), 559 (2023). https://doi.org/10.3390/sym15020559
26. Hatami-Marbini, A., Tavana, M.: An extension of the Electre I method for group decision-making under a fuzzy environment. Omega **39**(4), 373–386 (2011)
27. Chen, C.-T., Lin, C.-T., Huang, S.-F.: A fuzzy approach for supplier evaluation and selection in supply chain management. Int. J. Prod. Econ. **102**(2), 289–301 (2006)
28. Hofmann, F., Jaeger-Erben, M.: Organizational transition management of circular business model innovations. Bus. Strateg. Environ. **29**(6), 2770–2788 (2020)
29. Linder, M., Williander, M.: Circular business model innovation: inherent uncertainties. Bus. Strat. Environ. **26**(2), 182–196 (2015). https://doi.org/10.1002/bse.1906
30. Lacy and Partridge. Circular Economy Handbook. Palgrave Macmillan UK, [Place of publication not identified] (2020)
31. Vermunt, D.A., Negro, S.O., Verweij, P.A., Kuppens, D.V., Hekkert, M.P.: Exploring barriers to implementing different circular business models. J. Clean. Prod. **222**, 891–902 (2018)

32. Ertz, M., Leblanc-Proulx, S., Sarigöllü, E., Morin, V.: Advancing quantitative rigor in the circular economy literature: new methodology for product lifetime extension business models. Resourc. Conserv. Recycl. **150**, 104437 (2019)
33. Xingzhi, W., Yuchen, W., Ang, L.: Trust-driven vehicle product-service system: a blockchain approach. Procedia CIRP **93**, 593–598 (2020)
34. Schwanholz, J. and Leipold, S. Sharing for a circular economy? an analysis of digital sharing platforms' principles and business models. Journal of Cleaner Production, 269 (2020)
35. Tereschenko, O., Kieneke, T.: Erfolgsfaktoren: Stand der Forschung und Entwicklungsperspektiven. VDM Verlag Dr. Müller, Saarbrücken (2007)
36. Rösch, M.M.: Gießerei-Controlling: Erfolgsfaktoren von Gießereien und deren Steuerung. Fachverlag Schiele & Schön, Berlin (2013)
37. Bakker, C., Hollander, M. den, van Hinte, E., Zijstra, Y.: Products that last: product design for circular business models, TU Delft Library (2014)
38. Reim, W., Sjödin, D., Parida, V.: Circular business model implementation: a capability development case study from the manufacturing. Bus. Strateg. Environ. **30**(6), 2745–2757 (2021)
39. Selvefors, A., Rexfelt, O., Renström, S., Strömberg, H.: Use to use – a user perspective on product circularity. J. Clean. Prod. **223**, 1014–1028 (2019)
40. Whalen, C.J., Whalen, K.: Circular economy business models: a critical examination. J. Econ. Issues **54**(3), 628–643 (2020)
41. de Mattos, C., de Albuquerque, T.L.: Enabling factors and strategies for the transition toward a circular economy (CE). Sustainability **10**(12), 4628 (2018). https://doi.org/10.3390/su1012 4628
42. Morseletto, P.: Targets for a circular economy. Resourc. Conserv. Recycl. **153**, 104553 (2020). https://doi.org/10.1016/j.resconrec.2019.104553
43. Potting, J., Hekkert, M. P., Worrell, E., Hanemaaijer, A.: Circular economy: measuring innovation in the product chain. https://www.researchgate.net/publication/319314335_Circular_Economy_Measuring_innovation_in_the_product_chain
44. Aloini, D., Dulmin, R., Mininno, V., Stefanini, A., Zerbino, P.: Driving the transition to a circular economic model: a systematic review on drivers and critical success factors in circular economy. Sustainability **12**(24), 10672 (2020). https://doi.org/10.3390/su122410672
45. Goyal, S., Garg, D., Luthra, S.: Analyzing critical success factors to adopt sustainable consumption and production linked with circular economy. Environ. Dev. Sustainab. **24**(4), 5195–5224 (2021). https://doi.org/10.1007/s10668-021-01655-y
46. Lewandowski, M.: Designing the business models for circular economy—towards the conceptual framework. Sustainability **8**(1), 43 (2016). https://doi.org/10.3390/su8010043
47. Urbinati, A., Franzo, S., Chiaroni, D.: Enablers and barriers for circular business models: an empirical analysis in the Italian automotive industry. Sustain. Prod. Consum. **27**, 551–566 (2021)
48. Awan, U., Sroufe, R.: Sustainability in the circular economy: insights and dynamics of designing circular business models. Appl. Sci. **12**(3), 1521 (2022). https://doi.org/10.3390/app120 31521
49. Gusmerotti, N.M., Testa, F., Corsini, F., Pretner, G., Iraldo, F.: Drivers and approaches to the circular economy in manufacturing firms. J. Clean. Prod. **230**, 314–327 (2019)
50. Moktadir, A., Kumar, A., Mithun Ali, S., Kumar Paul, S., Sultana, R., Rezaei, J.: Critical success factors for a circular economy: Implications for business strategy and the environment. Bus. Strateg. Environ. **29**(8), 3611–3635 (2020)
51. Tura, N., Hanski, J., Ahola, T., Stahle, M., Piiparinen, S., Valkokari, P.: Unlocking circular business: a framework of barriers and drivers. J. Clean. Prod. **212**, 90–98 (2019)

52. Agyemang, M., Kusi-Sarpong, S., Khan, S.A., Mani, V., Rehman, S.T., Kusi-Sarpong, H.: Drivers and barriers to circular economy implementation: an explorative study in Pakistan's automobile industry. Manag. Decis. **57**(4), 971–994 (2019)

53. Chen, C.-W.: Improving circular economy business models: opportunities for business and innovation: a new framework for businesses to create a truly circular economy. Johnson Matthey Technol. Rev. **64**(1), 48–58 (2020). https://doi.org/10.1595/205651320X15710564 137538

54. Lahti, T., Wincent, J., Parida, V.: A definition and theoretical review of the circular economy, value creation, and sustainable business models: where are we now and where should research move in the future? Sustainability **10**(8), 2799 (2018). https://doi.org/10.3390/su10082799

55. Mentink, B. Circular Business Model Innovation: A process framework and a tool for business model innovation in a circular economy. http://resolver.tudelft.nl/uuid:c2554c91-8aaf-4fdd-91b7-4ca08e8ea621

56. Kuo, T.C.: Simulation of purchase or rental decision-making based on product service system. Int. J. Adv. Manuf. Technol. **52**, 1239–1249 (2010)

57. Hasanuzzaman, C.B.: Development of a framework for sustainable improvement in performance of coal mining operations. Clean Technol. Environ. Policy **21**(5), 1091–1113 (2019). https://doi.org/10.1007/s10098-019-01694-0

58. Roy, P., Tadele, D., Defersha, F., Misra, M., Mohanty, A.K.: Environmental and economic prospects of biomaterials in the automotive industry. Clean Technol. Environ. Policy **21**, 1535–1548 (2019)

59. Sharma, R.K., Singh, P.K., Singh, H.: A hybrid multi-criteria decision approach to analyze key factors affecting sustainability in supply chain networks of manufacturing organizations. Clean Technol. Environ. Policy **22**, 1871–1889 (2020)

60. Bocken, N.M., Harsch, A., Weissbrod, I.: Circular business models for the fastmoving consumer goods industry: desirability, feasibility, and viability. Sustain. Prod. Consum. **30**, 799–814 (2022)

61. Jabbour, C., et al.: Stakeholders, innovative business models for the circular economy and sustainable performance of firms in an emerging economy facing institutional voids. J. Environ. Manage. **264**, 110416 (2020). https://doi.org/10.1016/j.jenvman.2020.110416

62. Aranda-Usón, A., Portillo-Tarragona, P., Marín-Vinuesa, L., Scarpellini, S.: Financial resources for the circular economy: a perspective from businesses. Sustainability **11**(3), 888 (2019). https://doi.org/10.3390/su11030888

63. Rizos, V., et al.: Implementation of circular economy business models by small and medium-sized enterprises (SMEs): barriers and enablers. Sustainability **8**(11), 1212 (2016). https://doi.org/10.3390/su8111212

64. Vafaei, N., Ribeiro, R.A., Camarinha-Matos, L.M.: Selection of normalization technique for weighted average multi-criteria decision making. In: Camarinha-Matos, L.M., Adu-Kankam, K.O., Julashokri, M. (eds.) Technological Innovation for Resilient Systems. IAICT, vol. 521, pp. 43–52. Springer, Cham (2018). https://doi.org/10.1007/978-3-319-78574-5_4

65. Schwarz, E., Krajger, I., and Dummer, R. Innovationskompass für klein- und mittelständische Unternehmen: Neue Ideen finden und entwickeln. Linde Verlag GmbH (2006)

66. Owolabi, S.A., Mmereki, D., Baldwin, A., Li, B.: A comparative analysis of solid waste management in developed, developing and lesser developed countries. Environ. Technol. Rev. **5**(1), 120–141 (2016)

67. Thompson, A., Ny, H., Lindahl, P., Broman, G.I., Severinsson, M.: Benefits of a product service system approach for long-life products: the case of light tubes. In: 2nd CIRP International Conference on Industrial Product Service Systems (IPS2) (2010)

Digital Transformation for the Competitiveness of Export Companies Through Outsourcing, Chiclayo – Peru

Danna Johana Jiménez Boggio[✉] (iD), Mónica del Pilar Pintado Damián (iD), and Guillermo Segundo Miñan Olivos (iD)

Universidad Tecnológica del Perú, Chimbote, Perú
C20087@utp.edu.pe

Abstract. The objective of this article was to determine how digital transformation would allow optimizing the outsourcing of production processes in exporting companies, 2020–2021. Regarding the methodology, it is of an exploratory type applied and has a qualitative approach. Likewise, qualitative instruments such as documentary review, life history, focus groups, etc. were used. And the population included the export companies in the Lambayeque region that are active in the years 2020–2021. It was concluded that a digital transformation in the outsourcing of export companies must have friendly and intuitive processes, quick to handle, applicable for web and cell phone, an updated database that minimizes costs and times in MYPES in the sector.

Keywords: Outsourcing · digitalization · exporter · agro-export

1 Introduction

The continuous transformation that companies are currently experiencing requires dynamism, flexibility and efficiency; competitive success is generated through anticipating changes, focusing on the core business of the company and outsource all non-essential activities, using subcontracting with third-party companies specialized in the requested service, generating efficiency and incorporating external capabilities that it does not have, reducing costs and obtaining short-term results, this being a strategic business approach (Hidalgo, 2019). Therefore, outsourcing, as a business phenomenon, is gaining importance and attention among both scientists and entrepreneurs (Žitkienė and Dudė 2019).

Nowadays, information, as one of the most significant factors of production, is becoming a mega-important asset in any field of activity. This is conditioned by the importance of the alternative value of the information resource while being used in new business processes and for the realization of new business ideas (Paklina and Revenko 2019). The digitalization of the agricultural sector occurs with the application of digital technologies in different areas of activity, which allows the adaptation process to be

S.-H. Sheu (Ed.): IEIM 2024, CCIS 2070, pp. 204–214, 2024.
https://doi.org/10.1007/978-3-031-56373-7_16

completed more quickly and obtaining positive results (Tokareva et al. 2018; Ilchenko et al. 2019).

The use of virtual platforms within business processes allows for the optimization of activities and time, facilitating transactions, allowing their accessibility between entrepreneurs, suppliers and clients (Schwertner, 2017). Its use is crucial in business growth, keeping them at the forefront of technology, use of ICTs and articulating it to the actors involved (Proaño et al. 2018). Its use, effectiveness and importance must be evaluated to achieve the objective for which it was created (Ceballos et al. 2019).

The COVID-19 pandemic has challenged companies in various sectors of activity. Many of these organizations have been forced to adopt new internal working practices and have felt strong pressure to offer products through digital channels. Companies have experienced profound changes and in a very short time implemented solutions based on digital technologies (Uvarovau, 2021). At the same time, it has become necessary to redesign management and collaboration models to ensure that no one within organizations is left behind and feels excluded from this digitalization process (Almeida et al. 2020; Almeida, 2017; Verhoef et al., 2019).

Inevitably, organizations move down the path of digital transformation; However, a key question is whether they are ready for this change. Matt and Rauch (2020) and Pflaum and Gölzer (2018) indicate that companies, even those that are most advanced in the digital transformation of their activities and workflows, are not yet fully prepared to face the challenges of digital transformation. Digitalization requires a restructuring of processes, streamlining the company, investing in more organic structures, reinforcing standardization and automation, in order to optimize the ability to respond to customers.

Actions to mitigate the negative effects of COVID-19 involve the use of information systems supported by digital technology, to stay in touch and informed (Katz et al. 2020), making it necessary to address privacy and data protection issues. Within these platforms (García et al. 2018). Against this context, Spain designed a portal called Acelera Pyme, where companies obtain digital solutions (Agudelo et al. 2020).

Digitalization is vital for the survival of Mypes (Ulas, 2019), it allows them to use technology-based tools that confirm their post-pandemic survival and competitiveness (Esan connection, 2020; Moreira et al. 2018; Schallm et al., 2018). In this regard, aims to promote the digitalization of MSMEs, making use of the channels and network of contacts that the country has. Perú Marketplace is a tool where companies can publicize their products in a virtual and personalized way, they can make configurations such as profile, catalogues, audiovisual material, contact and brand. The participation of companies is free, as it is a state platform, registration and evaluation are sufficient (El Comercio, 2020).

The general objective of the research was: to propose digital transformation to strengthen the competitiveness of export companies through outsourcing, Chiclayo (Lambayeque) - Peru. Regarding the specific objectives, they are: Diagnose the outsourcing of production processes in exporting companies in the Lambayeque Department; identify the virtual tools used by exporting companies in the Lambayeque Region and define the design that the virtual outsourcing platform should have for exporting companies.

2 Methodology

The research is exploratory, Vara (2012) indicates that the qualitative exploratory design is essential when analyzing a poorly studied or new research problem. It helps the researcher to increase his knowledge and understand the problem and subsequently execute a better structured study. Qualitative instruments are used such as documentary review, life history, focus groups, unstructured observation, interviews, etc. (Escudero and Cortez 2018). With respect to the population, the following have been identified: Documents and reports from government institutions, Web pages. Scope: The research project has as its scope the agro-export companies in the Lambayeque region that are active in the years 2020–2021.

a) Companies that have facilities and provide agro-industrial production process services (service provider company).
b) Exporting companies in the agroindustrial sector that do not have a plant to process the exportable offer (service user company).

Population: No. of agro-export companies in the Lambayeque Region, updated as of 2021: 95 (Gercetur, 2021).

3 Results

3.1 Diagnosis of Outsourcing of Production Processes in Export Companies of the Lambayeque Department

It is evident in Fig. 1 that the total production 471, 484 TN is the installed capacity used, which constitutes 56% of the total installed productive capacity; Therefore, it is concluded that 841,936 TN is the total installed productive capacity of the exporting companies in the Lambayeque region.

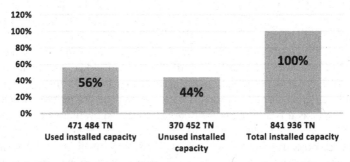

Fig. 1. Installed production capacity of exporting companies in the Lambayeque region, 2020

It is evident in Fig. 2 that 21% of export companies process pole beans, 13% handle mango, as well as those that process grapes, and 8% process avocado and coffee.

It can be seen in Fig. 3 that 89% of export companies that provide outsourcing services consider that their processing capacity per day is the main characteristic that users take into account.

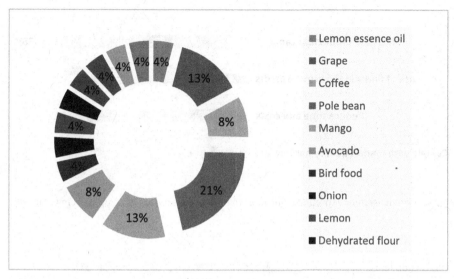

Fig. 2. Main products processed by exporting companies in the Lambayeque region, 2020

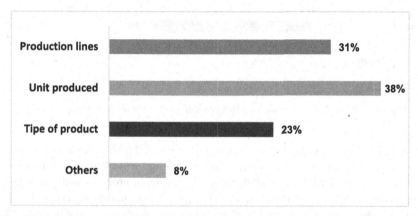

Fig. 3. Plant characteristics that are taken into account when contracting maquila services

Figure 4 shows that the main reason for exporting companies that led them to require the maquila services of outsourcing processing companies was to increase production capacity by 38%. It should be noted that the advantages of using the outsourcing service are access to productive technology by third parties, proximity to the port or point of departure of the product, and lower process costs.

Figure 5 shows that the export companies of Lambayeque that use outsourcing or maquila services pay for these for units produced at 38%, which are generally boxes, bags, cans of preserves or bottles.

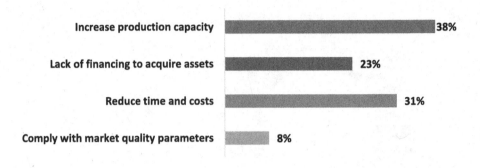

Fig. 4. Main reason for exporting companies to require the services of an outsourcing company

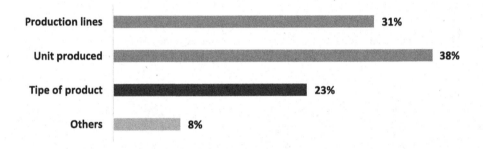

Fig. 5. Type of payment made by exporting companies for the outsourcing service

It is evident in Fig. 6 that 77% of exporting companies hire for the entire production process and 15% of companies hire only for packaging or packaging, 8% of companies hire for other services provided. However, the main problem with the maquila service provided by outsourcing companies is unpunctuality in the delivery of the product. In addition, companies receive fresh and processed products as final products from outsourcing companies. On the other hand, non-compliance with the contract by the maquiladora service provider is the most frequent risk faced by 50% of export companies that use these services, followed by the risk of theft of valuable information and adaptation to new technologies by of the service provider.

As seen in Fig. 7, 55% do have a website. Likewise, 64% of these have dynamic web pages; In addition, 56% of these have web pages, where Spanish and English predominate.

Figure 8 shows that 54% of companies that have a website do redirect to the company's social networks. In addition, 76% show the telephone number, email and form for direct contact. Likewise, 56% of them do have Facebook while only 20% have Instagram; Likewise, only 13% have Twitter. Finally, only 26% specify the product they export on their web pages.

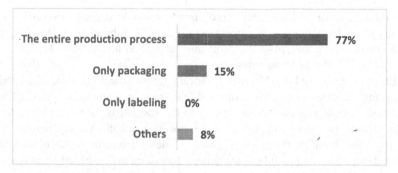

Fig. 6. Services that export companies contract from outsourcing companies

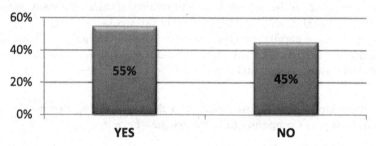

Fig. 7. Agro-export companies in Lambayeque that have a website – 2020

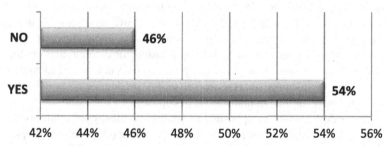

Fig. 8. Websites of the Agro-export companies of Lambayeque that redirect to social networks – 2020

3.2 Current Outsourcing Search Process for Agroindustrial Companies in Chiclayo - Peru

An interview was conducted with foreign trade experts and professionals who are part of agro-industrial export companies. According to this interview, the outsourcing selection process for exporting companies consists of several stages. Firstly, the commercial references of other exporting companies are taken into account. Then, the maquiladoras are visited to verify processes and facilities. It is important to consider the price per ton when selecting the outsourcing company and the ton per hour capacity in the process (installed capacity). In addition, compliance with certifications such as Global Gap, BRC, SENASA, HACCP, SMETA is verified. The cost/service ratio provided is also verified, as well as: infrastructure, cold chambers, storage capacity, experience time, proximity to production areas and access to paved roads. Subsequently, the companies make a comparison between all the companies consulted and visited. The cities and districts where exporting companies seek the maquila service are: Piura (Paita, Sullana), Lambayeque (Chiclayo, Motupe, Jayanca, Olmos), Ancash (Casma), La Libertad (Virú), Lima (Huaral, Cañete), Arequipa and Ica. The search and selection time for the maquiladora company is usually, according to the specialists, from one week to two including travel and sightseeing. Some of the specialists mentioned that there are companies that can take up to 30 days in the process. The approximate cost generated in the search and selection of a maquiladora company according to the specialists interviewed is: Visit to a plant, equivalent to a day of work including travel expenses, is quoted between 300 to 500 soles per person representing the exporting company. In a week for two people, which is the average human resource that makes the visits, we will be talking about S/ 4,000.00 soles and S/ 8,000.00 in two weeks.

For specialists, the starting point in the search for outsourcing is the recommendations and references received by other companies, which is why it is important to have visible comments from users in the digital transformation of this process.

3.3 Platform for Digital Transformation in the Outsourcing Process of Agro-Export Companies in Lambayeque - Peru

As mentioned, digital transformation is an extremely relevant process for every organization since it integrates digital technology in various business areas. Nowadays, it is necessary for organizations to adopt modern and innovative technologies to carry out cultural and operational changes that allow them to improve services to their clients or increase the productivity of their operations. In that sense, companies can design and implement digital solutions such as mobile applications or e-commerce platforms, in other cases they can migrate to a cloud computing infrastructure and in more operational matters they could apply smart sensors to reduce production costs; That is, digital transformation can be applied in many areas of companies.

Many small and medium-sized businesses already create digital solutions based on mobile applications or e-commerce platforms. A mobile application allows companies to design programs compatible with smart phones or tablets, which in turn increases the company's potential to transfer information from the company to its customers or suppliers. In the case of an e-commerce platform, there are online digital tools to sell products or services over the internet, however, internet platforms also have the ability to carry out internal management of the company with their respective suppliers, But many small and medium-sized companies miss this potential.

Digital transformation in the outsourcing process of agro-export companies in Lambayeque, Peru, can be extremely beneficial for efficiency and productivity. First of all, greater operational efficiency can be mentioned, since many services could be digitalized, allowing companies to concentrate on activities of greater relevance to the business, improving time management and reducing production and administrative costs. In the specific case of this study, it was determined that the hiring process of maquiladora companies in the agro-export sector is excessively face-to-face. By not having digital tools, flexibility and agility is lost in the process. Based on this problem, the use of resources and services increases to satisfy the company's needs. Improving this deficiency in Chiclayo's agro-export sector would be a first step to migrate to a digital transformation process and improve the competitiveness of the sector. An adequate strategic analysis would allow agro-export companies to digitalize the entire outsourced process to improve the choice of suppliers, control the quality of the process, apply business intelligence, among other alternatives. Likewise, supplier companies would have to go through a homologation process to adapt to the requirements of the agro-export sector, improving the quality of the service provided.

Currently, the outsourcing process requires a technical visit for each of the companies that could be included in the production process, however, the geographical location of each of them implies an extra cost of the process. The time involved can last two weeks and have an approximate cost between 4,000 and 8,000 soles, depending on the number of companies visited (Fig. 9).

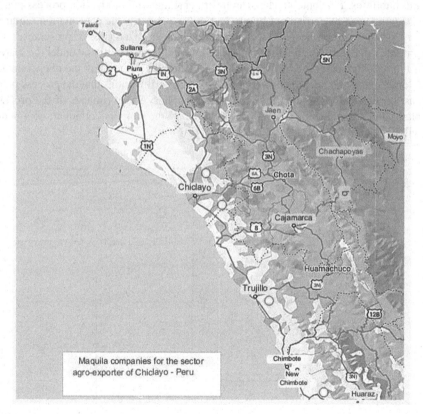

Fig. 9. Geographic location of outsourcing companies for the agro-export sector in Chiclayo - Peru

This study details the characteristics that a virtual platform should have for digital transformation in the outsourcing process of agro-export companies in Lambayeque – Peru. A virtual platform brings together user-friendly and intuitive processes, quick handling and downloading, App for web and cell phone, high resolution HD images, companies providing the outsourcing service may incorporate links to institutional videos, the service can be free or paid according to the complexity, updated database and description of the content of the virtual platform.

A) Registration: Both the service provider and the user must register on the platform to create their profile: The registration includes company information, company name, RUC, activity engaged in, name of legal representative, ID number, email, demonstrate medium risk level or better, operating license (for supplier companies). After validation of the data by the platform administrator, the username and access password are confirmed. Having verified that the companies are in a situation of assets and liabilities. B) Content: The search for agroindustrial production process plants (maquilas) can be carried out by product or by company: the presentation required for export must be selected: fresh, frozen or industrialized (juices, preserves, concentrates, dehydrated, among others). The information you will find under the search is the following: Name of the company – Outsourcing, location of the plant, contact (Name, email and telephone), services provided, certifications that the process plant has, service rates (optional), you will find a gallery space (images of the process plant) and comments on the service provided by the supplier (optional: comments will be made by other users of the service) (Fig. 10).

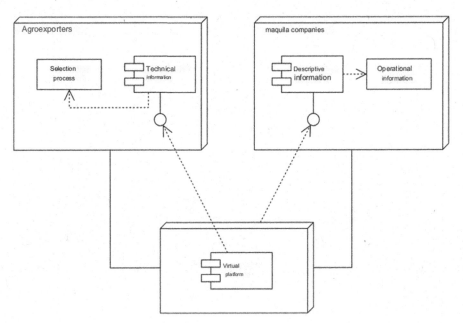

Fig. 10. Digital transformation model based on a virtual platform for agro-export companies in their outsourcing process

4 Conclusions

Only 65% of companies Exporters from the Department of Lambayeque have a processing plant own, of which 31% offer process outsourcing service or also is called maquila, within the which We find production, labeling and packaging. The users of this outsourcing consider by 89% the process capacity per day for the decision to contract the service and in 67% consider production lines, technology and certifications. Obtaining the service allows them to increase their production capacity, as well as reduce time and costs, obtaining access to technology from third parties who are located near the exit points of their goods. Payment for the service is made 38% per units produced, one of the main problems detected is unpunctuality in the delivery of the product and contracts are made per campaign in 69%.

55% of the export companies in the Department of Lambayeque have a website, 64% of these pages are dynamic, with response time measured in seconds. 56% are in Spanish and English, while only 2% consider other languages. 98% do not include a shopping cart due to the nature of marketing, 54% are redirected to social networks. Of the total agro-export companies, 56% have Facebook, 20% Instagram and 87% Twitter.

The digital transformation in the outsourcing process for export companies in Lambayeque must have friendly and intuitive processes, quick to handle, applicable for web and cell phone, and updated database. Containing: company name, location, contact, services, certifications, rates, images and comments.

References

Agudelo, M., et al.: Las Oportunidades de la Digitalización en América Latina frente al Covid-19, 1–33 (2020). https://n9.cl/jnmxv

Almeida, F.: Concept and dimensions of web 4.0. Int. J. Comput. Technol. **16**(7), 7040–7046 (2017). https://doi.org/10.24297/ijct.v16i7.6446

Almeida, F., et al.: Los desafíos y oportunidades en la digitalización de las empresas en un mundo post-COVID-19. en IEEE Engineering Management Review **48**(3) 97–103 (2020). https://n9.cl/qn3ok

Ceballos, O., et al.: Auditoría de usabilidad de herramientas implementadas en plataformas virtuales para ofertar servicios con responsabilidad social., Revista INNOVA ITFIP **5**(1), 64–77 (2019). https://n9.cl/mhtrp

Conexión Esan (12 de octubre del 2020). PYMES y la necesidad de digitalizarse para sobrevivir en la post pandemia. https://n9.cl/x0po2

El Comercio (25 de agosto del 2020). Mincetur lanza plataforma comercial para que exportadores peruanos vendan sus productos al mundo. https://n9.cl/x29ps

Escudero, C., Cortez, L.: Técnicas y métodos cualitativos para la investigación científica. Ediciones UTMACH (2018). https://n9.cl/bu9hq

García, J., et al.: Modelo de gestión de riesgos de seguridad de la información para PYMES peruanas. Revista Peruana de Computación y Sistemas **1**(1), 47–56 (2018). https://doi.org/10.15381/rpcs.v1i1.14856

Gercetur: Diagnóstico del Sector Exportador de la Región Lambayeque. Gerencia Regional de Comercio Exterior y Turismo del Gobierno Regional de Lambayeque (2019)

Hidalgo, A.: El outsourcing como estrategia competitiva de la empresa. Revista Cultural del Ateneo de Cádiz **19**, 1–16 (2019). https://n9.cl/wqknr

Katz, R., et al.: El Estado de la Digitalización de América Latina Frente a la Pandemia del COVID-19. CAF (2020). https://scioteca.caf.com/handle/123456789/1540

Matt. D., Rauch, A.: SME 4.0: The Role of Small- and Medium-Sized Enterprises in the Digital Transformation. Palgrave Macmillan, Cham (2020). https://doi.org/10.1007/978-3-030-254 25-4_1

Moreira, F., et al.: Enterprise 4.0 – the emerging digital transformed enterprise? Procedia Comput. Sci. **138**, 525–532 (2018). https://doi.org/10.1016/j.procs.2018.10.072

Paklina, L., Revenko, N.: Potentials of the use of IoT-technologies in agricultural sector Potentials of the use of IoT-technologies in agricultural sector. Adv. Intelligent Systems Res. **167**, 84–88 (2019). https://doi.org/10.2991/ispc-19.2019.19

Pflaum, A., Gölzer, P.: The IoT and digital transformation: toward the data-driven enterprise. IEEE Pervasive Comput. **17**(1), 87–91 (2018). https://doi.org/10.1109/MPRV.2018.011591066

Proaño, M., et al.: Los sistemas de información y su importancia en la transformación digital de la empresa actual. Revista espacios **39**(45) (2018). http://es.revistaespacios.com/a18v39n45/18394503.html

Schallmo, D.: Digital transformation of business models-best practice, enabler, and roadmap. Int. J. Innovation Manage. **21**(8) (2018). https://doi.org/10.1142/S136391961740014X

Schwertner, K.: Digital transformation of business. Trakia Journal of Sciences **15**(1), 388–393 (2017). https://n9.cl/q4c70

Tokareva. M., et al.: The influence of technologies of the Internet of things on the economics. Business Information Technologies (45), 6278 (2018). https://bijournal.hse.ru/data/2018/12/05/1144199880/6.pdf

Ulas, D.: Digital transformation process and SMEs. Procedia Computer Science **158**, 662–671 (2019). https://doi.org/10.1016/j.procs.2019.09.101

Uvarova, O.: SMEs digital transformation in the EaP countries during COVID. Eastern Partnership Civil Society Forum **19**(1), 1–68 (2021). https://n9.cl/ef63p

Vara, A.: Desde La Idea hasta la sustentación: Siete pasos para una tesis exitosa. Un método efectivo para las ciencias empresariales. Instituto de Investigación de la Facultad de Ciencias Administrativas y Recursos Humanos. Universidad de San Martín de Porres (2012). https://n9.cl/2h31

Žitkienė, R., Dudė, U.: The impact of outsourcing implementation on service companies. HAL science ouverte (2019). https://hal.archives-ouvertes.fr/hal-02167060/

A Shipborne Dense Storage Warehouse System Based on Two-Way Transportation Line Strategy

Miao He[1], Zailin Guan[1], Chuangjian Wang[2], and Guoxiang Hou[3]([⊠])

[1] State Key Lab of Digital Manufacturing Equipment and Technology, School of Mechanical Science and Engineering, Huazhong University of Science and Technology (HUST), Wuhan, China
[2] Key Laboratory of Metallurgical Equipment and Control Technology, Ministry of Education, Wuhan University of Science and Technology, Wuhan, China
[3] School of Naval Architecture and Ocean Engineering, Huazhong University of Science and Technology (HUST), Wuhan, China
D202080264@hust.edu.cn

Abstract. The shipborne warehouses store maintenance accessories, daily necessities, military equipment, etc., and they are important components of military support. These ships are mainly responsible for receiving, storing and distributing items. Due to limited ship space and increasing efficiency requirements for access operations, ship warehouses will gradually achieve automation, digitization, and intensive management. In this paper, the ship dense storage warehouse system greatly increases storage capacity. At the same time, the warehouse input and output operations are fast and accurate, shortening the operation time. The two-way transportation lines strategy is an improvement compared to the traditionally one-way strategy, and this two-way strategy is suitable for storing supplies which have wide varieties and large weights. The left and right horizontal movement lines increase the number of input and output ports, and the transportation tasks coordinated with AMRs make the transmission more flexible. The warehouse space is a whole space, and also can be several isolated spaces, which share elevators, and this space system is suitable for irregular, low-high, and multiple-layer spaces in ships.

Keywords: Shipborne dense storage warehouse system · Two-way transportation lines · Autonomous mobile robots (AMRs) · Sequential list

1 Introduction

In order to adapt to the multiple-purpose of modern large and medium-sized ships and ensure their material supply capacity during navigation, ships should have an integrated, compact, and highly automated warehousing system that maximizes storage capacity to store various types of supplies, such as maintenance accessories, daily necessities, military equipment, etc., as well efficient input and output management. The automated warehousing system enables automatic, reliable, accurate, and rapid selection, storage, and

S.-H. Sheu (Ed.): IEIM 2024, CCIS 2070, pp. 215–223, 2024.
https://doi.org/10.1007/978-3-031-56373-7_17

transportation of items. It not only improves operational efficiency, achieves unmanned operations, but also synchronizes logistics and information flow to achieve transparency and visualization of the processes. Furthermore, it can meet the timely need to improve the entry and exit of items without increasing personnel allocation. In the entire system, the storage and transmission modules are important subsystems for storage and transportation operations. Therefore, studying their storage strategy and transfer scheduling has great significance for the entire warehousing system.

2 Overall Researches for Automated Warehousing System

Shipborne warehouse storage allocation and transfer algorithm is less be concerned about than storage security [1–5], however, there are still some academic research on it. In [6] authors study the problem of ammunition transportation for major and medium caliber naval gun, aiming to realize high density storage and reduce labor intensity. They compare several ammunition storage schemes, such as library management mode, side-by-side ammunition clip and double channel mode, and propose an novel scheme named full-automatic magazine with high storage capacity and flexible changing and management. In [7] authors study demand uncertainties based on multiple-product inventory problem in a dual-channel warehouse, and establishing a DRMP inventory models to search for the optimal reorder points and order quantities for both the online and offline channels.

The basic principles for shipborne warehouse are as follows.

(1). Location management. It includes compartment storage [8], proximity storage [9, 10] and ensuring uniform force on shelves [11]. These studies improve the reliability of transportation [12], prevent damage, minimize the time for entry and exit, and improve the storage efficiency and facilitate management [13] according to the frequency and characteristics of ammunition.

(2). Job scheduling principles. There are many job scheduling principles, such as FIFO [14, 15], priority principle [16], shortest or longest processing time principle [17], random rule and assembling complete sets of components [18]. In [19] authors study combined picking and packing processes to reduce operation time and storage buffers, and present a generic algorithm consisted by container selection, loading configuration, and loading/picking sequence. These methods have their own advantages, for example, FIFO is suitable for similar tasks or priority systems that have minimal impact, and random rule is mainly considers the balanced response to each task.

(3). Transportation lines. Considering that the ships are often disturbed by waves and winds, and sometimes they are also disturbed by currents movements, the warehouses in ships should be sturdy and stable. The ship warehousing transmission module is designed as separation of horizontal and vertical movements. At the same time, the transportation lines in every layer are capable for moving the boxes horizontally to the elevators, and elevators are used for boxes vertical movement. Wherein the storage boxes are slotted on the transportation lines [6] to prevent slipping, and the slot design is easy to roll the boxes to elevators. Ensuring uniform force on belts, the storage box size is equal, and the bottom of boxes are toothed and fit with the gaps of the conveyor belts. When the belts are rolling, the boxes are moving along the transportation lines.

3   System Description and Problem Analysis

As the main combat unit at sea, ships store various types of supplies, such as equipment parts and foods. In order to ensure the requirements during daily life, it is necessary to coordinate and optimize the supporting tasks, reasonably arrange the scheduling order according by storage and transportation operations.

3.1   System Description

The storage and transportation system is responsible for the functions of items storage, retrieval and transportation. In our paper, it is composed of a warehouse, elevators, autonomous mobile robots (AMRs) for transferring items to the temporary storage area [20], as shown in Fig. 1 and 2.

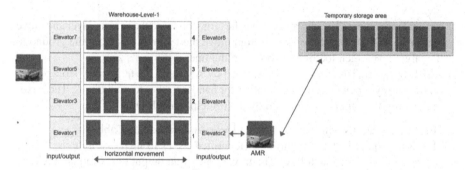

Fig. 1. The storage and transportation system (Warehouse-level-1)

In Fig. 1 and 2, there is an example of the dense storage and transportation system. The warehouse has several levels, and each level has 4 transportation lines, identified as 1, 2, 3 and 4. Each line has two elevators at both ends, which are used to loading or unloading boxes. Each line can move left or right independently according to the input/output tasks. Moreover, the elevators and AMRs work together to complete input and output tasks. This system improves the traditional one-way transportation [6] line movement, has a high feeding speed, simple control and transmission system, low manufacturing and installation accuracy requirements, and certain cushioning and shock absorption effects.

3.2   Model Description and Problem Analysis

To evaluate system performance, we define the scheduling processes as follows.

(1) Horizontal movement. The transportation lines are moving according by input/output tasks, and the target boxes are stopped at the end for elevator transport, or an empty unit is ready at the end for loading a box.

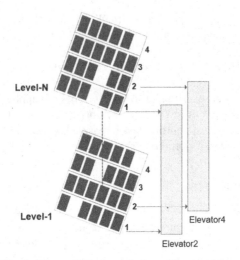

Fig. 2. The multiple layer design of the warehouse

(2) Vertical movement. The elevators move the target boxes to the bottom layer and move the boxes to the ground, or the elevators move the target boxes from the ground to the target level, then loading the boxes to the line.

(3) AMR transport. The boxes are transported by AMRs to the temporary storage area or the conveyer belt which is convenient for handing robot to grab boxes. The boxes in the temporary units are grabbed by the principle of FIFO.

There are notations will be used in the latter part, as shown in Table 1.

The time required to move the box (identified as x) from the warehouse to the temporary area is defined as follow. The travel time of an output task is equal, the time moving the target box to the left or right elevator, plus the waiting time for the elevator, plus the time moving the target box to the target level, plus the time waiting for the robots, plus the time moving the target box to the destination. The vertical space can be a whole space, also can be multiple isolated spaces.

$$Time_x = \frac{Length_{i,j,x}}{Speed_{line}} + Wait_{elevator} + \frac{Height_{level}\,(i-1)}{Speed_{elevator}} + Wait_{AMR} + \frac{Dis\tan ce_k}{Speed_{AMR}} \tag{1}$$

In formula 1, there are two waiting periods, such as waiting for an elevator to move the box from a certain line on a certain level to the ground, and waiting for a AMR to move the box from the ground to the temporary area. These waiting periods are the normal scheduling objectives aiming to minimize the travel time.

Under a series of items input and output instructions, we need to utilize limited resources (AMRs etc.) to complete more tasks, and our optimization goal is to minimize the total execution time and maximize equipment utilization.

Table 1. Table captions should be placed above the tables.

Label	Definition
Levels	The number of levels of the warehouse
Lines	The number of lines in each layer
Elevators	The number of elevators
AMRs	The number of AMRs
Temp	The number of temporary storage units
Length$_{line}$	The length of a line
Height$_{level}$	The height of a level
Length$_{i,j,x}$	The length of a box (identified as x) in i level, j line. i is [1, *Levels*], j is [1, *Lines*]
Speed$_{line}$	The speed of the line movement
Speed$_{elevator}$	The speed of the elevator movement
Speed$_{AMR}$	The speed of the AMR
k	The ID of the elevator, if the elevator is on the left side, $k = j$, and if the elevator is on the right side, $k = Lines + j$
Dis tan ce_k	The distance between the elevator-k to the temporary area, k is [1, *Elevators*]

4 Numerical Experiments

The supplies are stored in the warehouse by the following principles.

(1) The quantity of a certain type of item is multiple, and they are randomly stored in different layer or different transportation lines. This method increases the probability of successful output and avoids the probability of failure caused by transportation lines or elevators when a certain type of item is stored in a single line.

(2) The boxes located at both ends of the box sequence can be output by horizontal movement.

(3) When there are input tasks, they are stored in random empty locations in the warehouse. At the same time, when the elevator is suitable for performing double command (DC) operations, the input task can be combined with the output task.

Assuming a warehouse has 18 m, 9 layers and the storage capacity is 360. There are 5 transportation lines in each layer, and 8 storage units means 8 boxes can be stored in each line. Increasing the output quantity while changing the quantity of AMRs, the aim is to compare the equipment utilization and execution time. The output list is randomly generated, and the output tasks are executed in sequence. If some box cannot be output, the program is stopped. Then, we calculate the numbers of tasks have done. The procedure of the two-way transportation strategy is summarized as follow Algorithm 1.

Algorithm 1. The output tasks in warehouse based on two-way lines

---

Input: The quantity of item type is identified as $Type$, the horizontally movement is devided as $Left_H$ and $Right_H$, the quantity of output boxes is Q_{output}, and the probability of successful output operation is $P_{SuccessOutput}$ (Because the output tasks may be unsuccessful when the stored boxes on the ends are not matching the type of required in output list, or the matching boxes are not on the ends which can not be get out by elevators).

$Type$ is set from 11, 20, 25, 30 to 36.

Q_{output} is set from 20, 30, 40 to 50.

All storage units have been filled with boxes, and each type of items has equal probability and is randomly stored.

The output list is randomly generated, and the output tasks are executed in sequence. If some box cannot be moved out, the program is stopped. Then, we calculate the numbers of tasks have been done.

Procedure:

1:for output=1:OutputList
2:    for i=1:layers
3:        for j=1:(the quantity of lines per layer)
4:            for k=1:(the quantity of units per line)
5:                if (Type found on the left end) == output
6:                    ouput the box; break;
7:                elseif (Type found on the right end) == output
8:                    ouput the box; break;
9:                end
10:           end for
11:       end for
12:   end for
13:end for

At the same time, we use the MATLAB for multiple statistical calculations, as shown in Table 2.

From Table 2, the probabilities of successful output operations gradually decrease with the increase of number of output tasks and item types, from 100% to 6.5% in one-way, and from 100% to 42.34% in two-way. It's obvious that two-way method has greater success possibilities than one-way. The best case is when the number of item types is 11, and the number of output tasks is 20 in one-way. When the number of item types is 11 and 20, the number of output tasks is 20 to 40 in two-way. Which means the design of transportation line is suitable in this case. The worst case is when the number of item types is 36 and the number of output tasks is 50. Which means, the item types are various so that the target box cannot move out in sequence. Moreover, the reason

Table 2. The performance of output tasks

Type	Q_{output}	One-way	Two-way		
		$P_{SuccessOutput}$	$Left_H$	$Right_H$	$P_{SuccessOutput}$
11	20	93.1%	54.0%	46.0%	100%
11	30	87.2%	57.0%	43.0%	100%
11	40	86.8%	58.8%	41.2%	100%
20	20	57.9%	53.0%	47.0%	100%
20	30	41.2%	54.6%	45.4%	90.3%
20	40	36.8%	50.0%	50.0%	81.3%
25	20	34.4%	53.3%	46.7%	97.5%
25	30	23.4%	54.1%	45.9%	88.7%
25	40	19.2%	53.5%	46.5%	79.5%
30	20	22.4%	51.9%	48.1%	79.0%
30	30	19.6%	50.9%	49.1%	72.7%
30	40	11.7%	51.8%	48.2%	69.5%
36	20	14.8%	53.4%	46.6%	63.2%
36	30	9.7%	51.0%	49.0%	54.7%
36	40	7.1%	50.3%	49.7%	49.8%
36	50	6.5%	50.4%	49.6%	42.3%

that the left movements are slightly more than the right is due to the searching order is from left to right.

5 Conclusions and Further Research Opportunities

The shipborne dense storage warehousing system is based on an automatic warehouse system, providing an optimized input and output tasks cooperated with AMRs transportation method, which can effectively improve the support capacity of supplies. It greatly increases storage capacity compared with traditional ones. At the same time, the warehouse input and output operations are fast and accurate. The two-way transportation line strategy increases the number of input and output ports, greatly improves the efficiency of the whole warehouse system. More importantly, the warehouse space can be combined by several isolated spaces, and this space combination method is suitable for the irregular, low height and multiple-layer spaces in ships.

However, there are some disadvantages such as high cost (the elevators and AMRs are expensive), complex technical implementation, large system structure that need to be overcome. In the future, different item sizes storage problem, especially large size items in ships, will be considered.

Acknowledgement. We wish to thank the 5th International Conference on Industrial Engineering and Industrial Management for providing this chance to submit our paper. After that, thank you to all those who helped us during the paper writing.

References

1. Xu, Y., Shan, Y., Zhang, J., Fei, L.: Failure analysis in ammunition storage life modeling. In: 2014 Fourth International Conference on Instrumentation and Measurement, Computer, Communication and Control, Harbin, China, pp. 320–323 (2014). https://doi.org/10.1109/IMCCC.2014.73
2. Xiang, H., Wang, K., Li, Z.: Monitored control system of temperature/humidity for ammunition storehouse based on LabVIEW. In: 2011 First International Conference on Instrumentation, Measurement, Computer, Communication and Control, Beijing, China, pp. 172–175 (2011). https://doi.org/10.1109/IMCCC.2011.52
3. Liu, J., Ling, D., Wang, S.: Ammunition storage reliability forecasting based on radial basis function neural network. In: 2012 International Conference on Quality, Reliability, Risk, Maintenance, and Safety Engineering, Chengdu, China, pp. 599–602 (2012). https://doi.org/10.1109/ICQR2MSE.2012.6246305
4. Wang, Y.-W., et al.: Pressure relief of underground ammunition storage under missile accidental ignition. Defence Technol. **17**, 1081–1093 (2021)
5. Liu, J., An, Z., Zhang, Q., Zhao, T., Liu, G.: Research on safety assessment of gas environment in ammunition warehouse. In: 2013 International Conference on Quality, Reliability, Risk, Maintenance, and Safety Engineering (QR2MSE), Chengdu, China, pp. 1397–1399 (2013). https://doi.org/10.1109/QR2MSE.2013.6625830
6. Zhipeng, N., Hui, L., Ying, Z.: Research on ammunition storage overall technology of major and medium naval gun basing on AHP. In: 2016 International Conference on Robots & Intelligent System (ICRIS), ZhangJiaJie, China, pp. 316–319 (2016). https://doi.org/10.1109/ICRIS.2016.66
7. Qiua, R., et al.: A robust optimization approach for multi-product inventory management in a dual-channel warehouse under demand uncertainties. Omega **109**, 102591 (2022)
8. Torabizadeh, M., et al.: Identifying sustainable warehouse management system indicators and proposing new weighting method. J. Clean. Prod. **248**, 119190 (2020)
9. Sai Subrahmanya Tejesh, B., et al.: Warehouse inventory management system using IoT and open source framework. Alexandria Engineering J. **57**, 3817–3823 (2018)
10. Rebelo, C.G.S., et al.: The relevance of space analysis in warehouse management. In: 30th International Conference on Flexible Automation and Intelligent Manufacturing (FAIM2021) 15–18 June 2021, Athens, Greece
11. Brownlow, L.C., et al.: A multilayer network approach to vulnerability assessment for early-stage naval ship design programs. Ocean Eng. **225**, 108731 (2021)
12. Fichtinger, J., et al.: Assessing the environmental impact of integrated inventory and warehouse management. Int. J. Production Economics **170**, 717–729 (2015)
13. Son, D.W., et al.: Design of warehouse control system for real time management. IFAC-PapersOnLine 48(3), 1435–1439 (2015)
14. Alamri, A.A., et al.: Beyond LIFO and FIFO: exploring an allocation-in-fraction-out (AIFO) policy in a two-warehouse inventory model. Int. J. Prod. Econ. **206**, 33–45 (2018)
15. Lanza, G., et al.: Assigning and sequencing storage locations under a two level storage policy: optimization model and matheuristic approaches. Omega **108**, 102565 (2022)

16. Natalia Burganova, et al.: Optimalisation of internal logistics transport time through warehouse management: case study. In: 14th International scientific conference on sustainable, modern and safe transport, Transportation Research Procedia **55**, 553560 (2021)
17. Aravindaraj, K., et al.: A systematic literature review of integration of industry 4.0 and warehouse management to achieve Sustainable Development Goals (SDGs). Cleaner Logistics and Supply Chain **5**, 100072 (2022)
18. Peng, K., et al.: An effective hybrid algorithm for permutation flow shop scheduling problem with setup time. Procedia CIRP **72**, 1288–1292 (2018)
19. Shiau, J.-Y., et al.: A warehouse management system with sequential picking for multi-container deliveries. Comput. Ind. Eng. **58**, 382–392 (2010)
20. Rehman, S.: A parallel and silent emerging pandemic: antimicrobial resistance (AMR) amid COVID-19 pandemic. J. Infect. Public Health **16**, 611–617 (2023)

Application of the SLP Methodology to Improve Distribution in a Wood Furniture Factory

Nazira Alexandra Común Valle⦾, Luis Gerald Muncibay Rivas⦾,
Melany Kimberly Paitan Taipe(✉) ⦾, and Javier Romero Meneses⦾

Department of Industrial Engineering, Universidad Continental, Huancayo, Perú
72909232@continental.edu.pe

Abstract. The work was focused on a wood furniture manufacturing company, due to the growing demand for furniture, the company seeks to improve productivity to meet this demand and be competitive in the market.

In order to meet the growing demand, two alternative plant layouts were proposed using the SLP (Systematic Layout Planning) methodology; to increase production capacity distances were reduced. After analyzing several factors such as land availability, accessibility, costs, safety and more, it has been determined that alternative 1, which integrates all production areas on a single level, is the preferred choice due to its advantages in terms of efficiency and costs.

Finally, an economic analysis was performed with alternative 1 being the most convenient, giving us positive results, the financial analysis reveals a positive NPV (Net Present Value) of 584,815 Peruvian soles (Peruvian currency), indicating that the projected revenues exceed the initial costs, suggesting that the project can be profitable and beneficial for the company. This new plant design seeks to improve efficiency and productivity in the wood furniture industry in Peru.

Keywords: Systematic Layout Planning · Processes · Wooden furniture ·
Material flow · Design · Net present value

1 Introduction

Furniture production worldwide is in a constant process of innovation. More and more countries are increasing their production despite not having a wide range of raw materials. Global production of wood products in 2020 totaled 473 million m3 of sawnwood, 368 million m3 of wood-based panel [1].

The wood furniture manufacturing industry in Peru is increasingly improving and innovating in its production process. The furniture manufacturing industry in Peru is a sector that has a 78% share in the wood manufacturing market, being one of the most important in the second transformation production chain of the wood sector [2]. Observing the increase of it is economic activities in this sector, importance has been given to the production process in order to improve the productivity of the wood furniture industries and to advance at the same pace of the growth of the sector.

A company's productivity can be greatly affected by the design of its facilities. It is estimated that the cost of material flow contributes 30 to 70% of the total cost of

S.-H. Sheu (Ed.): IEIM 2024, CCIS 2070, pp. 224–236, 2024.
https://doi.org/10.1007/978-3-031-56373-7_18

manufacturing a product. Therefore, it is critical that the layout of facilities is organized effectively through a systematic methodology, also known as SLP [3].

Given that there is evidence of problems in the routes traveled by the operators, it is proposed to reduce distances to make better use of time in the main tasks [4]. In addition, by making the most of the spaces, the productive flow would change, allowing new working methods to be generated given the newly defined areas and establishing a new order in each one of them. This option would be especially effective and can be applied in the case of small and medium industries and/or companies, due to cost savings and time optimization of the production process.

Therefore, this research seeks to increase the production capacity of a wood furniture manufacturing company in Peru through the design of a new plant layout by comparing the proposed plant designs.

2 Theoretical Framework

2.1 SLP (Systematic Layout Planning)

This technique combines quantitative measurement of materials movement with non-flow considerations such as supervision, communications, personnel comfort and movement. It is major advantage being that it clearly documents the logic of the layout and easily allows input from all levels of staff [5]. That is, it consists of representing the plant organization in order to make the pertinent modifications to have a better plant layout.

2.2 Stages of SLP

- **Relationship analysis.** Performed after the processes for manufacturing the products have been established, the various functions within the company have been defined.
- **Relational table.** The successive columns are reduced in size until they disappear, leaving a triangular structure. Within the triangle there are cells that are divided, indicating the required proximity and the reason.
- **Relational path or activity diagram.** The path relational diagram is a graph in which areas are represented as nodes connected by lines. The latter represents the intensity of the relationship (A, E, I, O, U, X) given in the analysis of the relationships between activities according to the line code (A four lines, E three lines, I two lines, X dotted curved line). This details not only the type of operation that takes place in each area, but also allows analyzing the fact that the relationships type A, for the most part, are adjacent and those of type X are separated, thus fulfilling the objective of the SLP methodology [6].
- **Relational diagram of spaces.** Area Allocation Diagrams (AAD) shows templates globally, the information only shows the placement of area, while complete visualization images can be seen in the final template/layout which is the final result of analyzing and planning factory layout [7].
- **General layout as a whole.** The areas of the departments are reorganized in a plan, maintaining the measures of the initial proposal. The previously established relationships between the departments are also retained. In this way, the diagram is simplified and a clearer layout is shown.

- **Practical layout.** The areas are assigned according to the ideal layout to a plot plan divided into surface units. The space required for each area is respected, but the shape of the area can be adjusted, preserving the number of equivalent area units. This practical layout results in the final plan.

2.3 Redesign of Plant Layout

The optimum distribution is one that allows the material to cover the shortest possible distance between operations. You should always consider the distance traveled in each operation and choose the shortest, most comfortable and safest distance. It is a mistake to think that operations should not be successful [8].

3 Methodology

This section describes the phases that were carried out. The first phase consisted of the collection of the required data from the factory through visits to the factory, the problem has been considered in real time. The second phase was the application of the Systematic Layout Planning (SLP) method, starting with the planning, the design of the alternative layout with the SLP method is prepared based on the level of closeness that has been obtained from the Priority Scale Table (TSP) and Activity Relationship Diagram (ARD) [9]. The third phase was the discussion of the results obtained in order to determine the best option to grant a plant layout proposal.

4 Results

4.1 Description of the Process

The factors in a production plant are those that intervene in the productive process in a variable or susceptible way of variation and which alteration causes modifications in the result of the production process [10]. The production process starts from the cubing of planks to the storage of the furniture ready for sale. Figure 1 below shows the flow diagram of the solid wood furniture production process, which shows the sequences of activities.

4.2 Relationship Between Activities

The determination of the required area aims to design the layout of the proposed facilities that fits the needs of the production activities [11]. The layout problem is characterized by the selection of the best spatial positioning of facilities, such as equipment and workstations, in a specific area [12]. Having knowledge of the furniture making procedure and the flow of materials we can perform the analysis of relationships with the areas of the enterprise and the importance valued for each one. Proximity value: A (Absolutely necessary), E (Especially necessary), I (Importantly), O (Ordinary or secondary), U (Unnecessary).

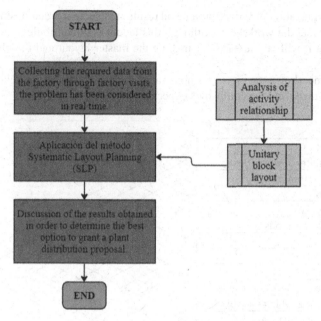

Fig. 1. Methodology flow chart

In Fig. 2, we can see that the most important relationship that now exists in the plant is that of the input warehouse with the boards, this may mean that for the current distribution, it is essential to have the raw material and its warehouse as close as possible, however, this does not follow an optimal flow with the other processes, given the fact that it must also follow a sequence to avoid too many comings and goings when transporting the product in process.

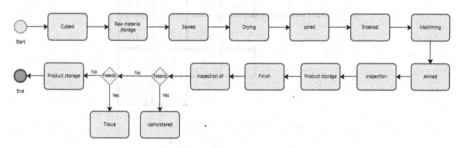

Fig. 2. Flow diagram of the furniture production process.

4.3 Relationship Between Areas

Once the closeness relationships between activities were determined, the next step consisted of proposing and evaluating the layout alternatives [13]. The dimensionless block diagram is the first attempt at distribution and result of the activity relationship chart and the worksheet and the worksheet. Although this layout is dimensionless, it will be the dimensionless, it will be the basis for making the master layout and the plan drawing [14].

The scheme shown in Fig. 3 demonstrates a clear coherence in relation to the modifications made and the company's expectations (Fig. 4).

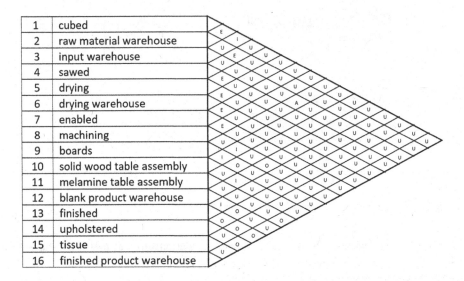

1	cubed
2	raw material warehouse
3	input warehouse
4	sawed
5	drying
6	drying warehouse
7	enabled
8	machining
9	boards
10	solid wood table assembly
11	melamine table assembly
12	blank product warehouse
13	finished
14	upholstered
15	tissue
16	finished product warehouse

Fig. 3. Relational diagram of the furniture manufacturing process.

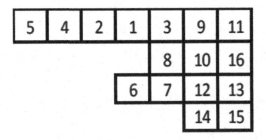

Fig. 4. Layout of unit blocks

4.4 Distribution Alternative 1

Alternative 1 consists of locating all production areas on the same level, favoring integration and communication between processes. To avoid dust contamination from wood furniture production, an air curtain is proposed in the melamine furniture production area. This air curtain prevents dust from spreading to other areas, improving air quality and the work environment. The air curtain can be regulated according to production needs, ensuring a safer and healthier working environment for employees.

4.5 Distribution Alternative 2

This alternative offers a differential advantage by having a second level exclusively for melamine products, which are protected from the dust generated in the manufacture of wooden furniture. For this purpose, filtration, ventilation, cleaning and maintenance systems are used to ensure a clean and suitable environment for storage. In addition, the company has its own elevator for the transfer of products, which allows for safe, fast and efficient transport. This solution guarantees the care and quality of the melamine products, as well as the optimization of the company's internal logistics processes.

4.6 Evaluation of the Proposed Alternatives

The existing relationships between each pair of activity centers should not be limited to the intensity of the transport intensity of the flow of materials or people between them, because sometimes these relationships are irrelevant or nonexistent. This is why the qualitative relationships that may exist between the activity centers that justify the adjacency requirements or their undesirability must also be quantified [15].

- **Qualitative method.** The factors to be evaluated are the following: Land availability (1), plant access (2), labor availability (3), proximity to market (4), safety (5), transportation costs (6), land cost (7), implementation cost (8), plant size (9) (Tables 1 and 2).

Table 1. Qualitative method

FACTOR	1	2	3	4	5	6	7	8	9	NUMBER	WEIGHTED (%)
1		1					1			2	5.56%
2	1		1		1		1	1		5	13.89%
3					1					1	2.78%
4	1	1	1		1	1	1	1	1	8	22.22%
5	1							1		2	5.56%
6	1	1	1		1		1	1		6	16.67%
7		1		1					1	3	8.33%
8	1		1				1		1	4	11.11%
9	1	1	1		1	1				5	13.89%
TOTALS										36	100%

According to the analysis, alternative 1 proved to be the most appropriate, since it offers greater optimization of space, better distribution of resources, less exposure to

Table 2. Qualitative method

FACTOR	WEIGHTED (%)	PROPOSAL 1		PROPOSAL 2	
		Qualification	Score	Qualification	Score
Land availability	5.56%	6	0.333333	4	0.222222
Plant access	13.89%	8	1.111111	8	1.111111
Labour	2.78%	6	0.166667	6	0.166667
Closeness to the market	22.22%	8	1.777778	8	1.777778
Security	5.56%	8	0.444444	6	0.333333
Transportation costs	16.67%	8	1.333333	8	1.333333
Land cost	8.33%	6	0.5	6	0.5
Implementation cost	11.11%	6	0.666667	2	0.222222
Plant size	13.89%	6	0.833333	8	1.111111
TOTALS	100%		7.166667		6.777778

occupational hazards and greater staff motivation. Therefore, it is recommended that alternative 2 be implemented as the definitive solution to the problem.

4.7 Comparison of Proposed Alternatives

• **Factor analysis.** After having carried out the qualitative methods, an analysis of factors such as investment, meters, advantages and disadvantages, etc. of both alternatives is carried out.

Table 3. Analysis of Factor of alternatives

	Units	Alternative 1	Alternative 2
Advantages and Disadvantages	Points	60	48
Factors	Points	2.27	1.91
Investment	S/	4,269,902	4,277,243
Meters traveled per unit produced	m	295	450
N° of Weighted Setbacks	N°	0	0.11

Based on what is shown in Table 3, Alternative 1 is the best option for the company due to it is qualitative and quantitative advantages in the integration between work areas, it is lower investment cost and its higher efficiency in terms of distance traveled and setbacks, which boosts production.

4.8   Economic Evaluation

The economic evaluation allows us to observe the feasibility of the project (Table 4).

Table 4. Economic analysis of alternative 1

	Economic Cash Flow							
	Year 0	Year 1	Year 2	Year 3	Year 4	Year 5	Year 6	Year 7
Income								
Income from increase of production				464,597	464,597	464,597	464,597	464,597
Factory sales income current production/ Machinery	1,288,980		2,130,346					
Income from investment savings by eliminating transport between plants				57,000	57,000	57,000	57,000	57,000
Income from investment savings reduce costs storage				107,065	107,065	107,065	107,065	107,065
Investment income of others savings				8,400	8,400	8,400	8,400	8,400
Total revenues	1,288,980		2,130,346	637,062	637,062	637,062	637,062	637,062
Expenses								
Initial investment for implement the project	3,846,123	303,979	119,800					
Operators cost				294,680	294,680	294,680	294,680	294,680
Maintenance cost of new machinery				1,200	1,200	1,200	1,200	1,200
Total expenses	3,846,123	303,979	119,800	295,880	295,880	295,880	295,880	295,880
Effective flow	-2,557,143	-303,979	2,010,546	341,182	341,182	341,182	341,182	341,182
Discount rate (COK)	2.35%	Annual						
VAN FCE	584,815							
TIR FCE	9%							
VAN Benefits	6,274,667							

(continued)

Table 4. (*continued*)

| | Economic Cash Flow | | | | | | | |
	Year 0	Year 1	Year 2	Year 3	Year 4	Year 5	Year 6	Year 7
VAN Costs	5,575,379							
B/C	1.1							
PERIOD OF RECOVERY (YEARS)	6							

The financial analysis indicates that the project has a positive Net Present Value (NPV) of 584,815 Peruvian soles. This suggests that, in present value terms, the project's revenues exceed the initial costs and expenses. This is an indication that the project has the potential to generate profits and add value to the company. Furthermore, the investment will be recovered in the sixth year.

4.9 Layout Alternative 1

In relation to the qualitative analysis, alternative 1 was shown to be the most versatile, favoring greater integration between work areas and providing additional advantages by facilitating effective communication and coordination between them (Figs. 5 and 6).

- **First floor**

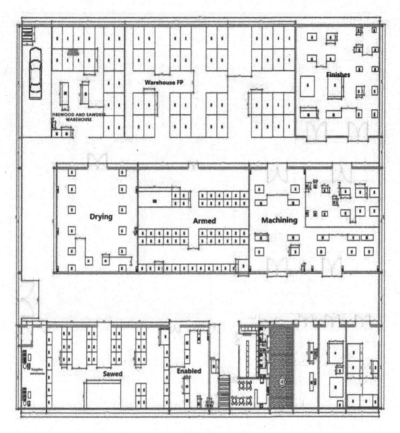

Fig. 5. Layout of first floor alternative 1

- **Second Floor**

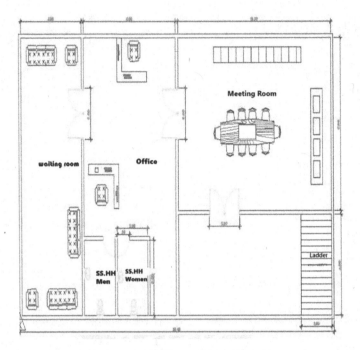

Fig. 6. Layout of second floor alternative 1

5   Discussion

Companies that do not perform the plant layout exercises are exposed to low material flow efficiencies because their layouts may not be in accordance because they organize their areas without performing the adjacency analyses required in the SLP methodology [6]. Alternative 1 is the best option for the company because it has a second level dedicated to melamine products, which avoids damage from wood dust. In addition, it has an elevator to move materials and products to the second level. Alternative 2, on the other hand, has everything on one level and uses an air curtain to separate the wood and melamine production areas. Alternative 1 is more economical, more integrated and more efficient than Alternative 2. For these reasons, Alternative 1 is recommended as the final solution.

6   Conclusions

- The redistribution design of the furniture production plant is the best option for the company because it avoids the saturation of its current physical spaces and infrastructure, which was only conditioned for the factory.

- Job satisfaction is benefited through the proposal, since it has a dining room, spacious restrooms, showers, gardens, and ventilated spaces, among others.
- In this work, two plant layout alternatives were compared for a manufacturing company dedicated to the production of wooden furniture, using the qualitative method.
- Alternative 1 allows a better organization of the flow of materials and processes, greater ease of control and supervision, less exposure to occupational hazards and greater ability to adapt to changes in demand or product design.

References

1. International Forestry - Forest Research. Forest Research [en línea]. 29 de septiembre de 2022 [consultado el 19 de septiembre de 2023]. https://www.forestresearch.gov.uk/tools-and-resour ces/statistics/forestry-statistics/forestry-statistics-2022/9-2/
2. Ministerio de la Producción (PRODUCE) (2020): Estadística Comercio Interno, Lima (PRODUCE)
3. Lista, A.P., et al.: Lean layout design: a case study applied to the textile industry. Production [en línea]. **31** (2021). ISSN 1980–5411. https://doi.org/10.1590/0103-6513.20210090
4. Palominos, P., Pertuzé, D., Quezada, L., Sanchez, L.: An extension of the systematic layout planning system using QFD: its application to service oriented physical distribution. Eng. Manag. J. **31**(4), 284–302 (2019). https://doi.org/10.1080/10429247.2019.1651444
5. Khariwal, S., Kumar, P., Bhandari, M.: Layout improvement of railway workshop using systematic layout planning (SLP) – a case study. Materials Today: Proceedings [en línea]. ISSN 2214–7853. https://doi.org/10.1016/j.matpr.2020.10.444
6. Torres, K.J., et al.: Metodología SLP para la distribución en planta de empresas productoras de Guadua Laminada Encolada (G.L.G). Ingeniería [en línea] **25**(2), 103–116 (2020). ISSN 2344–8393. https://doi.org/10.14483/23448393.15378
7. Gozali, L., Widodo, L., Nasution, S.R., Lim, N.: Planning the new factory layout of pt hartekprima listrindo using systematic layout planning (SLP) method. IOP Conference Series: Materials Science and Eng. **847**, 012001 (2020). https://doi.org/10.1088/1757-899x/847/1/ 012001
8. Ortiz Naranjo, E.J., Zúñiga Valle, A.X.: Distribución de planta y sus factores: incidencia en el mejoramiento de la productividad. Revista de Investigaciones en Enèrgía. Medio Ambiente y Tecnología: RIEMAT ISSN: 2588-0721 [en línea]. **7**(1), (2022) [consultado el 18 de septiembre de 2023]. ISSN 2588-0721. https://doi.org/10.33936/riemat.v7i1.4840
9. Haryanto, A.T., Hisjam, M., Yew, W.K.: Redesign of facilities layout using systematic layout planning (SLP) on manufacturing company: a case study. IOP Conference Series: Materials Science and Engineering [en línea] **1096**(1), 012026 (2021). [consultado el 17 de septiembre de 2023]. ISSN 1757-899X. https://doi.org/10.1088/1757-899x/1096/1/012026
10. Alpala, L.O., et al.: Methodology for the design and simulation of industrial facilities and production systems based on a modular approach in an "industry 4.0" context. DYNA [en línea] **85**(207), 243–252 (2018). ISSN 2346–2183. https://doi.org/10.15446/dyna.v85n207. 68545
11. Suhardi, B., et al.: Facility layout improvement in sewing department with systematic layout planning and ergonomics approach. Cogent Engineering [en línea]. **6**(1), 1597412 (2019). ISSN 2331-1916. https://doi.org/10.1080/23311916.2019.1597412
12. Benitez, G.B., et al.: Systematic layout planning of a radiology reporting area to optimize radiologists' performance. J. Digital Imaging [en línea] **31**(2), 193–200 (2017). [consultado el 18 de septiembre de 2023]. ISSN 1618-727X. https://doi.org/10.1007/s10278-017-0036-9

13. Pérez-Cerquera, M.R., Hurtado Londoño, J.A., Cruz-Bohorquez, J.M.: Applied electromagnetics course with a conceiving-designing-implementing-operating approach in engineering education. Ingenieria y Universidad [en línea] **26**, 1–32 (2022). ISSN 2011-2769. https://doi.org/10.11144/javeriana.iued26.aecw

14. Arias Morales, J.A.: Diseño del Proceso Productivo Para Una Nueva Linea de Cervezas Artesanales. Universidad Católica Andrés Bello (2019)

15. Gosende, P.A.P.: Evaluación de la distribución espacial de plantas industriales mediante un índice de desempeño. Revista de Administração de Empresas [en línea] **56**(5), 533–546 (2016). ISSN 0034–7590. https://doi.org/10.1590/s0034-759020160507

Author Index

S.-H. Sheu (Ed.): IEIM 2024, CCIS 2070, pp. 237–238, 2024.
https://doi.org/10.1007/978-3-031-56373-7

Printed in the United States
by Baker & Taylor Publisher Services